ASPHALTENE DEPOSITION CONTROL BY CHEMICAL INHIBITORS

Theoretical and Practical Prospects

ALI GHAMARTALE

PhD Candidate at the Department of Process Engineering, Memorial University, St. John's, NL, Canada

SHOKUFE AFZALI

PhD Candidate, Department of Process Engineering, Memorial University, St. John's, NL, Canada

NIMA REZAEI

Assistant Professor in the Department of Separation Science, at the Lappeenranta-Lahti University of Technology, Lappeenranta, Finland

SOHRAB ZENDEHBOUDI

Associate Professor and Equinor Chair in the Department of Process Engineering at Memorial University, St. John's, NL, Canada

ELSEVIER

G P P

Gulf Pro

An imp

Gulf Professional Publishing is an imprint of Elsevier
50 Hampshire Street, 5th Floor, Cambridge, MA 02139, United States
The Boulevard, Langford Lane, Kidlington, Oxford, OX5 1GB, United Kingdom

Library of Congress Cataloging-in-Publication Data
A catalog record for this book is available from the Library of Congress

British Library Cataloguing-in-Publication Data
A catalogue record for this book is available from the British Library

ISBN: 978-0-323-90510-7

For information on all Gulf Professional publications
visit our website at https://www.elsevier.com/books-and-journals

Publisher: Joe Hayton
Senior Acquisitions Editor: Katie Hammon
Editorial Project Manager: Chris Hockaday
Production Project Manager: Prasanna Kalyanaraman
Cover Designer: Miles Hitchen

Typeset by SPi Global, India

ASPHALTENE DEPOSITION CONTROL BY CHEMICAL INHIBITORS

Contents

CHAPTER ONE

Asphaltene and asphaltene precipitation/deposition

1.1 Introduction

Asphaltenes are the heaviest, and the most polar among the crude oil constituents [1]. They are defined by the solubility criteria such as species that are soluble in aromatic solvents (e.g., toluene, benzene, or pyridine) but insoluble in alkanes (e.g., *n*-pentane or *n*-heptane) [2, 3]. The asphaltenes have over 10^5 different types of molecular structures [4]. Even if asphaltenes of different oils have a comparable solubility, their chemical properties can be completely different [5]. Reservoirs with a lighter crude oil are being depleted faster than heavy oil and bitumen hydrocarbon resources, and the use of heavy oil and bitumen are becoming more common in refineries. Accordingly, the asphaltene problems have become more pronounced as the mentioned unconventional resources provide feedstock to the industrial refining systems [6, 7]. The precipitation and deposition of asphaltene in the production and transportation of crude oil cause two main problems: formation damage, and plugging of wellbore and production equipment [8, 9]. The complications from asphaltene stability and the necessity to support oil banks have remarkable harmful side effects on the whole production chain, such as:

- Production impairment by altering the reservoir wettability and pore plugging [10–13].
- Depositions of the asphaltene in the well [14] and refining facilities [15].
- Arterial clogging of the destabilized asphaltene particles in the pipelines and wellbores [16, 17], leading to the formation of water and oil emulsions [2].
- Contamination and corrosion of production facilities and catalysts [7, 15, 18, 19].
- Deactivation of catalyst in hydrocarbon processing [20].
- Creation of coke [15, 21–23].

Asphaltene Deposition Control by Chemical Inhibitors
https://doi.org/10.1016/B978-0-323-90510-7.00006-9

Asphaltene conservation and asphaltene deposition rejection are considered as approaches that overcome the economic, environmental, and technical problems due to asphaltene flow-assurance issues [2]. In the conservation approach, the asphaltene content of the crude oil is inhibited from precipitating by the addition of solubilization agents [24–29], a dilutant [16, 30], or stabilized oils [31]. In the asphaltene rejection approach, the inhibition of asphaltene deposition is conducted by solvent deasphalting [7, 26, 32]; solvent deasphalting is performed by emulsion extraction [33, 34], ultrafiltration [35], fluid carbon rejection [36], and selective oxidation of heteroatom-containing species followed by their selective separation [37]. Various experimental methods are used to detect the asphaltene deposition onset at reservoir conditions. These methods include a gravimetric technique, acoustic resonance technology (ART), light scattering technique (LST), high-pressure microscopy (HPM), high-pressure and high-temperature (HPHT) filtration, and quartz crystal resonance (QCR). Some of these techniques, such as LST and HPM, become industry standards. Nevertheless, considering the high costs associated with these methods, most of the current research studies are conducted at ambient pressure, using dead oil, and thermodynamic models are implemented to evaluate the asphaltene behavior at reservoir conditions [38].

A comprehensive thermodynamic knowledge of asphaltene precipitation and deposition is needed as the first step in planning processes to reduce the asphaltene problems. Thermodynamic principles govern the equilibrium distribution of the acting components in different phases. For crude oil systems, the thermodynamic studies can predict if the asphaltenes are stable or instable (including a second solid phase that can be potentially deposited onto surface facilities) [38]. The deposition of asphaltene in pipeline or wellbore can substantially reduce the productivity of an oilfield; in extreme cases, the pipeline or well can be completely plugged. Currently, huge investments are annually carried out to prevent, mitigate, or dissolve the asphaltene precipitates and deposits. Extensive research studies are also conducted to develop cost-effective inhibition or mitigation approaches of asphaltene deposition. Various strategies are devised to prevent asphaltene deposition in the wellbore, including asphaltene chemical inhibitors, controlling the operating conditions (pressure, temperature, and flow rate), and providing an internal surface coating. There are different asphaltene inhibitors, designed to stabilize asphaltenes by reducing the aggregate size, but their performances are related to oils and asphaltene characteristics. Despite all efforts to understand the asphaltenes and control asphaltenes' adverse effects, asphaltene deposition remains a challenge for the industry and more research and engineering activities are needed to find comprehensive and cost-effective treatment ways.

1.2 Asphaltenes: Properties, structures, and thermodynamic behaviors

The term *Asphaltene* was initially used by J.B. Boussingault in 1837 for the residue compounds of the bitumen distillations, which are insoluble in alcohol and soluble in turpentine [39]. This initial definition is similar in some aspects to the current asphaltene description, which argues that asphaltene has insoluble components in *n*-alkanes, such as *n*-heptane, but is soluble in toluene [16]. Asphaltenes are a branch of hydrocarbon, which have a dark color, containing heavy compounds with an average density of 1.2 g/cm^3, and they are fragile and infusible solids [40]. They do not feature a unique specific melting point and leave a carbonaceous residue while being decomposed on heating. Asphaltenes significantly increase the viscosity of crude oil due to self-association [41]; the high viscosity impedes oil transportation and reduces the yield of petroleum distillates due to high resistance against cracking [42]. Asphaltenes are undesirable from an oil waste management perspective due to having heavy metals and consequently the lack of biodegradability [43]. Nevertheless, asphaltenes have applications in industry [44, 45], such as surface protective coatings (in road construction) because of their adhesion properties, bio-inertness, and rheological toughness [42]. They also incorporate in drilling fluids to control fluid loss.

Elemental analysis of asphaltenes shows that they are mostly composed of carbon and hydrogen with an average C/H ratio of 1–1.2 [16]. In addition to carbon and hydrogen, asphaltenes contain *heteroatoms* such as nickel, sulfur, nitrogen, oxygen, and vanadium. In the literature, two general types of chemical structures are proposed for the asphaltene molecules. The first one is referred to "continental or island" form [46, 47] that is formed by a single polycyclic aromatic hydrocarbon (PAH), being attached to pendant alkyl groups. The second structure is "archipelago" form, wherein multiple PAHs (usually two) are linked together by alkyl groups [48, 49] (Fig. 1.1). These two structures are distinguishable unless the continental structure asphaltenes aggregate and form a structure similar to the archipelago structure; this might be a reason for the prevalence of the archipelago molecular structure idea for asphaltene structure in the open sources [50,51]. Recently, the results of asphaltene disaggregation along with the laser adsorption and

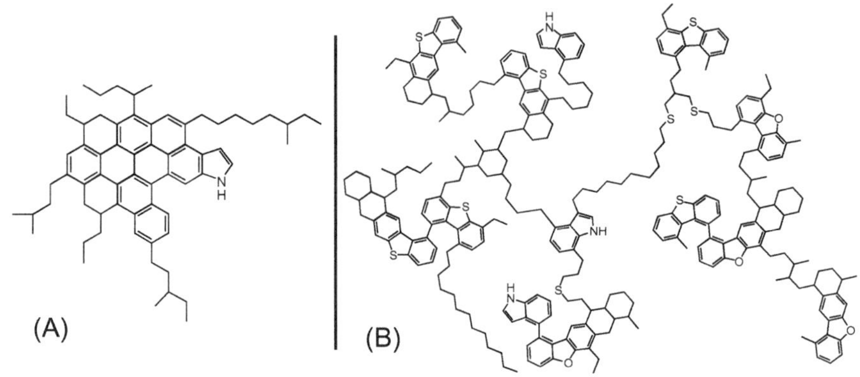

Fig. 1.1 Common chemical structure types of asphaltenes: (A) continental/island model [47] and (B) archipelago model [49].

atomic force microscopic tests have confirmed that the continental structure is the dominant structure for asphaltene [52,53].

Scanning tunneling microscopy (STM) provides a unique analytical methodology in imaging single adsorbates at an atomic scale. The asphaltene molecular structure can be characterized using STM [54,55]. In recent developments, atomic force microscopy (AFM) enables the visualization of the atomic structure of individual molecules on real surfaces [56]. This technique is also applied to analyze the bond order [57], obtain the molecular structure of natural compounds [58,59] and graphene nanoribbons [60], and to identify chemical synthesis products [61] and surface reactions [62,63]. In addition, the molecular orbitals can be mapped using STM and ultra-thin insulating films as a substrate [64,65].

The characterization of asphaltenes based on their solubility is more challenging than their chemical classes compared to lighter components of hydrocarbons. The lighter fractions of hydrocarbon, such as saturates and aromatics, are easily categorized according to their chemical structures while providing such distinctions for the heavier fraction of crude oil is difficult. In other words, the lighter fractions of crude oil have much simpler chemical structures and better economic perspectives in the industry; also, they are easier to measure and analyze than the heavy components [16]. For instance, gas chromatography characterizes light hydrocarbon components (<C36), while more complicated methods are required to extract detailed data about the structures of heavy components. Some techniques are extensively applied in the examination of asphaltene and other heavy compounds, such as mass spectrometry, electron microscopy, nuclear magnetic

resonance, small-angle neutron, X-ray scattering, ultrasonic spectroscopy, dynamic light scattering, fluorescence depolarization, vapor pressure osmometry, and gel permeation chromatography. Table 1.1 summarizes the recent works on asphaltene characterization [20,66].

Table 1.1 Summary of recent works about characterization of asphaltenes[a] [20].

Author	Technique	Objective
Wong and Yen [66]	ESR	To examine the ability of microwave power to dissociate petroleum asphaltenes macrostructures.
Leon et al. [67]	ST, EC, VPO, ^{13}C NMR	To study aspects related to the asphaltene deposition and composition of crude oils.
Joshi et al. [68]	OST	To study the characteristics of the asphaltene precipitate.
Shirokoff et al. [69]	XRD	To report the structural characterization of Saudi Arabia asphaltenes.
Sharma et al. [70]	HRTEM	To test the stacking and stacking disruption.
Buenrostro et al. [71]	FD, FES, ^{13}C NMR, IR	To explore the aspects of asphaltene chemical structure.
Alboudwarej et al. [72]	VPO	To examine the effect of different degrees of washing on the asphaltene properties.
Yarranton et al. [74]	VPO, IT	To study the molar mass and solvent-water interfacial tensions of asphaltenes.
Gutierrez et al. [73]	VPO, TLC	To isolate asphaltene fractions and study the solubility in toluene.
Guiliano et al. [75]	GC/MS, EC, FTIR	To investigate the potential of supercritical fluid extraction for crude oil asphaltenes.
Groenzin and Mullins [76]	FD	To survey the molecular size of a broad range of asphaltenes.
Fuhr et al. [77]	TGC	To examine the analytical separation of asphaltenes, wax, and soluble oil fractions.
Domin et al. [78]	SEC, PDMS, VPO, LDMS, MALDI	To examine and compare the molecular-weight distributions.
Groenzin and Mullins [46]	FD	To analyze FD rates of very dilute solutions of asphaltenes.

Continued

Table 1.1 Summary of recent works about characterization of asphaltenes—cont'd

Author	Technique	Objective
Carbognani et al. [79]	SEC, IR, ^{13}C NMR	To study the precipitation of asphaltenes during oil production and storage.
Artok et al. [80]	Py/GC/MS, GPC, 1H/^{13}C NMR, MALDI TOF	To elucidate the distribution of the aliphatic carbons.
Peramanu et al. [81]	VPO, GPC, EC, ^{13}C NMR	To analyze the molecular weight of asphaltenes and their composition.
Tojima et al. [82]	VPO, ^{1}H NMR	To apply a new asphaltene fractionation method using a binary solvent system.
Miller et al. [83]	VPO, SEC, EC, SANS, ^{13}C NMR, LDMS, HRMS	To separate and characterize Mayan asphaltene in two fractions.
McLean and Kilpatrick [84]	FTIR, ^{13}C NMR	To determine and compare the composition of asphaltenes of four different crude types.
Andersen [85]	FTIR, EC, HPLC-SEC	To investigate whether asphaltenes could be extracted directly from the crude oil by a polar solvent mixture.
Andersen [85,86]	1H/^{13}C NMR, FS, HPLC-SEC, VPO, EC	To investigate the effects of the temperature on precipitation of asphaltenes.
Fuhr et al. [87]	VPO, ^{13}C NMR	To study some properties of asphaltenes precipitated as a function of the solvent and temperature.
Adarme et al. [88]	VPO, ^{13}C NMR	To study how metals are removed from asphaltenes and study thermal and catalytic effects.

[a]OST: Optical scattering technique. FD: Fluorescence depolarization. FES: Fluorescence emission spectroscopy. ^{13}C NMR: 13-Carbon Nuclear magnetic resonance. IR: Infrared. ST: Surface tension. EC: Elemental composition. VPO: Vapor pressure osmometry. GPC: Gel permeation chromatography. IT: Interfacial tension. ESR: Electron spin resonance. GC/MS: Gas chromatography/mass spectrometry. FTIR: Fourier transformed infrared. SEC: Size-exclusion chromatography. HTGC: High-temperature gas chromatography. XRD: X-ray diffraction. py/GC/MS: Pyrolysis gas chromatography/mass spectrometry. MALDI TOF: Matrix-assisted laser desorption/ionization time of flight. PDMS: Plasma desorption mass spectrometry. LDMS: Laser desorption mass spectrometry. HRTEM: High-resolution transmission electron microscopy. SANS: Small-angle neutron scattering. HRMS: High-resolution mass spectrometry. TLC: Thin-layer chromatography. FS: Fluorescence spectroscopy. HPLC: High-performance liquid chromatography.

Asphaltenes have the highest fraction of metals among other constituents of crude oils [89]. The metals existing in the asphaltene structure are in the form of metalloporphyrins, and heterocyclic macrocycle organic compounds that interconnect with internal ligands and form complex structures. The metal content also exists at the defect center of the aromatic structures [89].

Two conventional methods for analyzing the metal content of the asphaltenes are atomic adsorption and mass spectroscopy. In both methods, the asphaltene sample should be free from organic components. To eliminate the organic part, a heat treatment technique such as calcination is implemented, following an acidic digestion process, using concentrated (70 wt%) nitric acid [38]. Fourier transform infrared (FTIR) spectroscopy is also used to determine the asphaltene chemical structure; this analytical method identifies the organic and inorganic compounds according to the absorption of the mid-IR radiation in the wavelength range of 2.5 μm (4000 cm^{-1}) and 25 μm (400 cm^{-1}) of the light spectrum [38]. As the asphaltene sample is exposed to the IR radiation, different bonds of molecules absorb a specific radiation wavelength, shifting their dipole moment and also the state of excited vibration [38]. The absorbed light wavelength of each bond is controlled by the energy difference of the base and excited vibrational state, which links the absorbed light wavelength to the molecular structure of the compound. The output of the FTIR test is a plot of radiation absorbance (or transmittance) against the light wavenumber (cm^{-1}) that is the reciprocal of the wavelength [90]. Table 1.2 provides the main adsorption peaks for the common components in crude oil.

Table 1.2 Absorption bands (FTIR) for different components of crude oil [38].

Functional group	Chemical structure	Absorption bands, FTIR (cm^{-1})
Alkanes and cycloalkanes	CH_3-CH_2	2950–2850 1470–1350 725–720
Aromatics	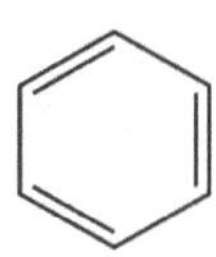	3080–3030 1625–1575 1525–1450
Pyrrolic nitrogen	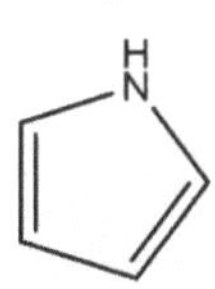 	3500–3400 1590–1560 1540–1500
Pyridinic nitrogen	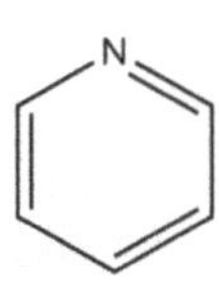 	3450–3200 1610–1360

Continued

Table 1.2 Absorption bands (FTIR) for different components of crude oil [38]—cont'd

Functional group	Chemical structure	Absorption bands, FTIR (cm^{-1})
Alcohols	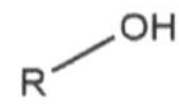	3550–3200 1250–970
Ketones	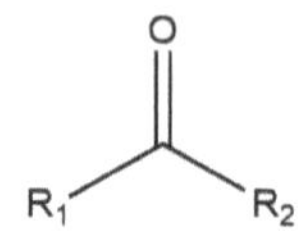	1750–1680
Quinones	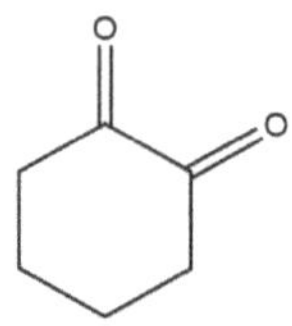	1650–1630
Ethers	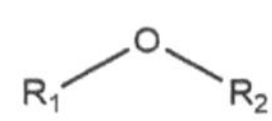	1150–1085
Esters	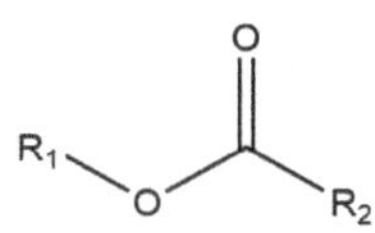	1750–1735
Carboxylic acids	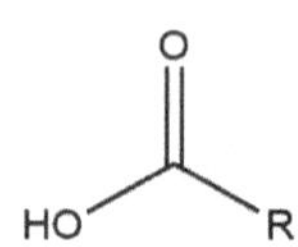	3550–2500 1780–1710
Sulfoxides	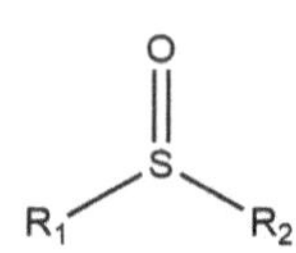	1225–980
Thiphenes	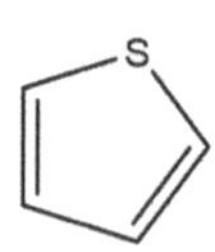	710–570
Sulfides	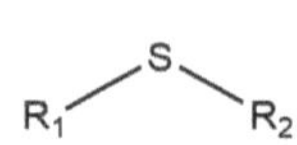	710–570

Nuclear magnetic resonance (NMR) spectroscopy also characterizes the chemical structure of the asphaltene based on the atoms' nuclear spin. A magnetic field is created as a result of the nuclear spin in the presence of an external magnetic field [38]. The energy applied for the magnetic field is related to the energy difference between the aligned and misaligned nucleuses, and electromagnetic signals are emitted at specific frequencies according to the energy difference. The emitted signals are collected and reported on a graph of intensity versus frequency (or chemical shift, in ppm). The most conventionally analyzed nuclei using NMR include 1H, 13C, 15 N, and 31P; among these tests, 1H and 13C are commonly used to characterize the asphaltenes. The oil industry employs the NMR method to compute the aliphatic and aromatic carbon fractions as well as the alkyl-substituted and unsubstituted aromatic carbons in a given asphaltene sample [29,90,91].

1.3 Asphaltene precipitation/deposition: Description, mechanisms, and field conditions

Asphaltene compounds are initially stable in crude oil, but they flocculate, precipitate, and deposit due to temperature, pressure, and composition changes in the system [92]. Flocculation is the asphaltene molecules' agglomeration. Precipitation is a solid-phase formation in the liquid phase (while Hirschberg et al. [93] defined the asphaltene precipitation as a liquid-liquid separation), and deposition occurs when the solid particulates are placed onto a surface [94].

The asphaltene flocculation and precipitation can be described with the micellization theory [95]. Micellization is the self-association of the surface-active materials in a liquid phase. The asphaltene monomer is soluble in oil; however, the asphaltene aggregates are usually not present in the form of monomers due to their surface activity and the presence of precipitant substances in crude oil (such as saturates). Asphaltenes form aggregates as a result of self-aggregation in the presence of paraffins (Fig. 1.2). Nevertheless, resins are also present in crude oil and the aggregates can cause steric colloids in the presence of an excessive amount of resins and paraffins (Fig. 1.3). Asphaltene micelles are formed if there is an excessive amount of aromatic substances in crude oil (Fig. 1.4). Therefore, based on the micellization theory, if the asphaltene concentration exceeds the critical micelle concentration (CMC)

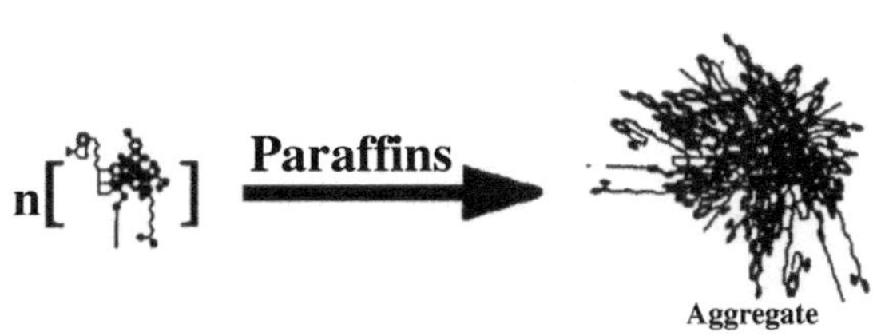

Fig. 1.2 Flocculation of asphaltenes and formation of asphaltene aggregates [95].

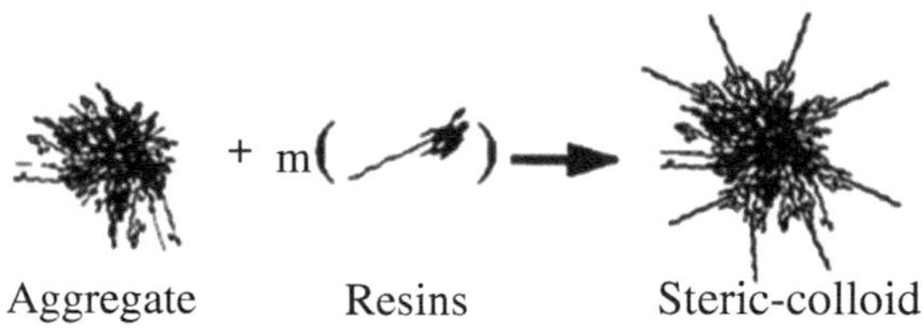

Fig. 1.3 Formation of asphaltene steric colloid [95].

level, asphaltene micelles will form. Increasing the asphaltene concentration causes growing micelles followed by forming coacervates and asphaltene phase separation [96]. Mousavi-Dehghani et al. [96] claimed that the average size of asphaltene molecules is 5 nm, the average size of asphaltene micelles is 25 nm, and the average size of micellar coacervates is greater than 25 nm.

Although the experimental evidence confirms the asphaltene micellization phenomenon, the theoretical research studies on this subject are rare or sparse. Nevertheless, extensive studies investigate the thermodynamics theory of phase separation in micellar solutions for oil-surfactant-water system [95,97,98]. Originally, this theory was introduced for water-amphiphile solutions, and later, it was extended to be employed in investigations on phase separation of aromatic asphaltene micellar solutions [95].

The Yen-Mullin model [99] is another approach to define the asphaltene flocculation and deposition. In this model, the aggregation is considered as a stepwise process [99]: the first step is nano-aggregation formation as a result of asphaltene agglomeration; the second step is cluster formation from self-associated nano-aggregates, which is followed by flocculation of the asphaltene molecules; and in the final step, sphase separation occurs, and asphaltene aggregates precipitate and deposit in the medium. Numerous studies are conducted to obtain the average size of these aggregates where the approximate range of 2–3 nm is reported for nano-aggregations, the average size of 5 nm is reported for the clusters, and the flocculate size is 300 nm to 1 μm [99].

One of the controversial subjects in asphaltene studies is the reversibility of the asphaltene precipitation. Pfeiffer and Saal [100] confirmed the irreversibility of the asphaltene precipitation process through an experimental study.

Fig. 1.4 Formation of asphaltene disk-like, cylindrical, and spherical micelles [95].

However, Hirschberg et al. [93] showed the reversibility of the precipitated asphaltene particles through a thermodynamic model. Kokal et al. [101] observed that by adding the heavy oil sample, the precipitated asphaltenes of two different crude oil samples redissolved whether using an aromatic solvent or not. Chaisoontornyotin et al. [102] demonstrated that the asphaltene precipitation process is reversible by changing the temperature. They also explained that using a precipitant along with temperature change can result in an irreversible asphaltene precipitation phenomenon. Their results suggested that there are two competing asphaltene precipitation mechanisms and, thereby, the precipitant-induced aggregation needs to be isolated [102]. Vargas et al. [103] studied the induced asphaltene precipitation in the presence of *iso*-octane (iC_8) using optical microscopy. They observed that the addition of iC_8 caused the asphaltene aggregation and when the iC_8 was evaporated from the solution, the asphaltenes dissolved into the oil.

The formation of asphaltene aggregates causes a problem in the extraction, transport, storage, and refining processes of asphaltic crude oil, and the elimination of them in hydrocarbon production and processing stages drastically increases the operating costs [104]. Crude oils with low API and high asphaltene content do not necessarily create asphaltene deposition problems [105]. The main parameter in asphaltene aggregation is related to asphaltene stability rather than the asphaltene content. For instance, a heavy oil reservoir with a high asphaltene content (in the case of asphaltene stability at reservoir condition) is less probable to experience asphaltene problems than the light oil reservoir with the same asphaltene content. The asphaltene precipitation issue is common in light crude oils with the well pressure above the bubble point [16, 106]. In addition, undersaturated reservoirs are more likely to experience asphaltene precipitation or deposition problems [16]. Fig. 1.5 shows a schematic of a phase envelope for a crude oil along with asphaltene-precipitation envelope (APE). In fact, the APE (blue dot curve) defines the instability zones for the asphaltenes in solution. At a given reservoir condition (red dot), pressure decreases due to the primary reservoir depletion. When the pressure reaches the upper asphaltene-precipitation envelope, also known as the asphaltene-precipitation onset pressure, the least soluble asphaltenes will precipitate. As pressure continues to decrease, more asphaltenes will precipitate until the bubble point pressure is reached. If the pressure is further decreased, asphaltenes are re-dissolved until the lower asphaltene-precipitation envelope is reached. The redissolution is due to less portion of light components after gas liberation [16]. The aggregation of asphaltene usually starts at the approximate concentration range

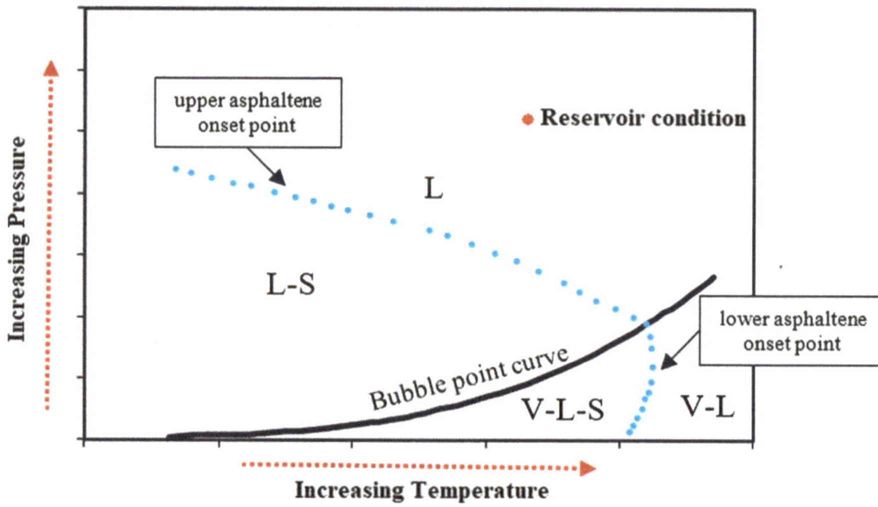

Fig. 1.5 Asphaltene-precipitation envelope (APE) in a pressure-temperature plot [109].

of 50–100 mg/L and depends on different parameters, such as temperature, pressure, solvent power of oil matrix, and the existence of other surface-active particles in the system [74,99,107,108].

The mechanisms that are commonly considered responsible for the asphaltene self-aggregation are: dipole-dipole interactions (due to hetero-atoms), and induced dipole-induced dipole interactions (due to resonance bond); and the induced dipole-induced dipole interactions that occur in three configurations, including, (1) face-to-face, (2) T-shape, and (3) offset π-stacked [47, 110]. Recently, Gray et al. [111] stated that the Brønsted acid-base interactions, hydrogen bonding, metal coordination complexes, and interactions between cycloalkyl and alkyl groups (to form hydrophobic pockets) are also responsible forces in the asphaltene aggregation.

de Boer et al. [112] developed an asphaltene supersaturation plot called "de Boer plot" (Fig. 1.6) as a simple approach to screen crude oils in terms of the asphaltene deposition at field conditions. The plot is generated through correlating the crude oil properties such as solubility, molar volume, and asphaltene solubility in the oil versus the density of the crude oil under in situ conditions. The *y*-axis in Fig. 1.6 is the difference between the reservoir pressure and bubble point pressure, and the *x*-axis is the insitu crude oil density.

Based on the de Boer plot (and the reservoir data), it is possible to assess the risk level of asphaltene deposition at various production conditions.

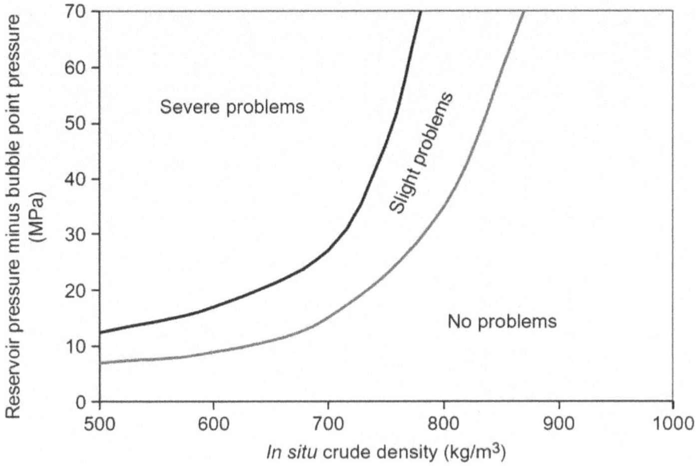

Fig. 1.6 de Boer crude oil supersaturation plot. *(Based on de R. De Boer et al., Screening of crude oils for asphalt precipitation: theory, practice, and the selection of inhibitors, SPE Prod. Facil. 10 (01) (1995) 55–61.)*

According to the field data, de Boer et al. concluded that the asphaltene deposition is more prone to happen in the reservoir and transportation system when crude oil is light with C_1-C_3 (>37 mol%), with relatively low C_{7+} content (<46 mol%), high bubble point pressure (>10 MPa), and a high difference between reservoir and bubble point pressure [112]. In such crude oils, the amount of asphaltene content is usually low (<0.5 wt%). In heavy crude oils, the percentage of components to suspend asphaltenes, such as resins, are higher compared with precipitant components, such as saturates. This difference becomes more significant when the pressure approaches the bubble pressure. Therefore, heavy crude oils with high asphaltene content are less likely to cause deposition problems, especially when the reservoir pressure is close to the bubble point pressure.

1.4 Modeling asphaltene precipitation and deposition

Extensive research is conducted on the modeling of asphaltene precipitation and deposition. In addition, artificial intelligence and statistical theory are applied to optimize the model tuning parameters, and to overcome problems such as nonlinearity and uncertainties that are pertinent in asphaltene precipitation and deposition systems. In this section, we present a variety of models for asphaltene precipitation and deposition in brief;

more information is provided in Chapter 6, which is dedicated to modeling of the asphaltene inhibitor. The thermodynamic-based models for asphaltene precipitation are categorized into thermodynamic colloidal model, regular solution model, equation of state (EoS) models, polymer solubility models, solid models, and thermodynamic micellization theory.

1.4.1 Thermodynamic colloidal model

Leontaritis and Mansoori [113] were the first researchers who proposed the thermodynamic colloidal model. This model is based on the assumption of moderate dissolution of the asphaltenes in the crude oil that act as a colloidal suspension. In this model, resins are considered to be absorbed at the surfaces of the asphaltene particles, resulting in the stability of the suspended particles. The asphaltene precipitation onset is predicted by determining the critical chemical potential for the resins, using the Flory-Huggins model. Several laboratory observations considered the asphaltene association to be irreversible, which contradicts the hypothesis of the thermodynamic colloidal model. Therefore, the colloidal model is more applicable to the dissociation of the asphaltene.

1.4.2 Regular solution model

The regular solution model is based on the Gibbs free energy which relates to enthalpy and entropy. In this model, it is assumed that these two terms can be treated separately [3]. This model considers a liquid-liquid equilibrium so that the heavy liquid phase contains asphaltene and resin (asphaltene-rich phase), and the lighter phase includes saturates and aromatic components (solvent-rich phase) [114]. The regular solution model needs mole fraction, molar volume, and solubility for each component as the input parameters [115].

1.4.3 EoS models

The simplest approach to estimate the asphaltene precipitation is by applying EoS approach in which the heavy fractions of the crude oil are presented as *semi*component groups of precipitating and nonprecipitating cuts [116, 117]. Considering the asphaltene precipitate as a pure solid constituent of the crude oil, the EoS models use three-phase flash calculations to estimate the asphaltene precipitation. The micellar character of the asphaltene is ignored in this method.

1.4.4 Polymer solution theory

In this technique, asphaltene mixture (or crude oil) is regarded as a nonideal solution mixture. This model combines the conventional solution theory with the Flory-Huggins polymer solutions theory [118] to estimate the maximum volume fractions of the asphaltene dissolved in the crude oil. In this model, a vapor-liquid equilibrium flash calculation is performed, using an EoS to obtain the bulk properties and equilibrium compositions of the vapor and liquid phases. Next, a liquid-liquid equilibrium calculation is performed by applying the Flory-Huggins theory to obtain the asphaltene fraction. Burke et al. [119], Kokal et al. [120], and Novosad and Costain [121] performed experimental studies to enhance the accuracy of the existing solubility models. The micellar character of the asphaltene is notconsidered in the polymer solution theory. The main drawback of the solubility model is the need for extensive data points to obtain the model parameters. The validity of these models strongly depends on the reversibility of the asphaltene precipitation.

1.4.5 Solid models

In the solid models, the asphaltene is considered as a solid component, and the model parameters are tuned by fitting to the experimental results. The solid model was initially proposed by Chung et al. [122], where the heavy fractions of the crude oil are classified into two pseudo-components: precipitating (e.g., asphaltenes) and nonprecipitating (e.g., resins). Given the simplicity of this model, it is capable of identifying the asphaltene stability in the crude oil. One drawback of the solid model is that the effect of pressure on the asphaltene stability onset is not considered.

1.4.6 Thermodynamic micellization model

According to the micellization model [123], the asphaltenes are assumed to aggregate and form a core that is surrounded by other oil components and resins. The addition of light normal alkanes to the crude oil results in the precipitation of asphaltene monomers. In this method, the growth rate of asphaltene micelles and the extent of the precipitation are estimated through minimization of the Gibbs free energy. This model shows promising results when compared with experimental data. The model has been tested with both live and dead oils for light crudes.

1.4.7 Other models

Lack of thermodynamic models on prediction of the asphaltene phase behavior encouraged researchers to develop novel and accurate models to address previous models' drawbacks. Rassamdana et al. [124] used the aggregation-gelation phenomena using the scaling-fractal theory to derive a simple scaling equation. In this new model, the amount of precipitated asphaltene (wt%) was defined as a function of two parameters: dilution ratio (R) and the molecular weight of the injected fluid or diluent (Mw). Subsequently, Hu et al. [125] assessed Rassamdana et al.'s model and verified the accuracy and simplicity of the model for different oils; they also found optimal values for the model exponents. In addition, the effects of temperature and pressure were added to the original model [124, 126]. Bagheri et al. [127] added the effects of gas-to-oil ratio, resin-to-asphaltene ratio, mole fraction of heavy components, molecular weight of residual oil, and oil specific gravity. Hence, they suggested a modified scaling equation for the prediction of asphaltene precipitation at a variety of dilution ratios; however, the model was limited to ambient pressure and temperature [128, 129]. Scaling equations and thermodynamic models are generally difficult, computationally demanding, and time consuming. Unconventional and hybrid approaches transform the modeling techniques related to the asphaltene phase behavior. Because the conventional methods generally cannot overcome the complexity of processes or phenomena and lead to error, researchers are attracted to predictive modeling techniques, which are less costly, time saving, and also accurate in application for extremely nonlinear systems. In recent years, smart techniques, such as Bayesian belief network and artificial neural network with various optimization algorithms including imperialist competitive algorithm, genetic algorithm, and particle swarm optimization, have been applied to predict asphaltene precipitation or deposition processes with high accuracy based on collective data bank available in the open sources [94,130–135].

Modeling of asphaltene deposition is a step after the modeling of asphaltene precipitation. Asphaltene deposition can be studied from two different perspectives: thermodynamics and transport phenomena. According to the thermodynamic aspect, the unstable thermodynamic precipitation conditions are identified. Two main classifications are used in asphaltene thermodynamic modeling. In the first modeling strategy, asphaltenes and resins are considered to have molecular entities; in the second modeling strategy, asphaltenes and resins are suspended aggregates in the crude oil medium.

From the transport phenomena prospect, the mass fraction dynamics of the deposited asphaltene is determined. The fraction of asphaltene that is deposited on the surface of the transport media (bulk scale) is a function of temperature, flow velocity, and mass diffusion of asphaltene in the porous media and/or pipelines. The mass transfer and adsorption coefficients of asphaltene depend on the asphaltene particle size. Therefore, deriving a model for asphaltene particle development requires simultaneous analysis of thermodynamic equilibrium and transport phenomena.

Two mathematical models that are commonly used to predict the asphaltene deposition behavior in porous media are adsorption and mechanical entrapment . The "adsorption" model assumes that the asphaltene particles attach to the rock pore walls as a result of polarity and intermolecular forces. This adsorption leads to a reversible reduction in the asphaltene concentration in the crude oil. Various studies have shown that asphaltene adsorption in the porous media follows a Langmuir isotherm type [136–138]. Although the asphaltene adsorption has a minor influence on the rock porosity, it drastically changes the rock wettability and fluid permeability. In the "mechanical entrapment" model, empirical correlations are used to predict the fluid flow path and pore blockage in the porous media as a result of asphaltene deposition. This model was developed based on the Groesbeck and Collins [139] theory for particle migration and deposition in porous media. According to this theory, the asphaltene deposition, permeability reduction, and path blockage are studied in a system with two parallel flow paths corresponding to the plugging and nonplugging pathways, while they have different aperture diameters and asphaltene particle sizes. The principal model parameters are the distribution of rock pore throat, pore channel, and asphaltene particle size. Asphaltene deposition modeling in EOR/IOR involves various thermodynamics, kinetics, and transport phenomena aspects that need to be comprehended through theoretical and experimental investigations.

1.5 Theoretical and practical challenges in asphaltene deposition

The deposition of asphaltene is a global issue that occurs in the majority of oil fields, such as in Venezuela, the Persian Gulf, Adriatic Sea, Mexican Gulf, and Alberta [140]. Wettability alteration is one of the undesired consequences of asphaltene deposition (from water-wet to oil-wet) for reservoirs' rock. Among various subcategories of the crude oil, resin and asphaltene remarkably affect the wettability of the reservoir rock. Asphaltene effects are much pronounced in alternating the wettability of rock surface due to the high polarity associated with their functional groups and higher propensity towards aggregation [10].

During oil production, various reservoir properties, such as pressure, temperature, and composition, change, which can cause asphaltene precipitation and deposition. These issues become more intense when the heavy and extra heavy oil reserves become the primary source for refining sectors [6, 7]. The complexities associated with the asphaltene instability adversely affect the production chain; indeed, wells are affected at first by decreased rate of oil production due to changes in the wettability, pore plugging [10–13], and asphaltene deposition in wells [14].

1.6 Economic analysis and environmental prospects of asphaltene deposition

The destructive effects of asphaltene deposition are extensively discussed in the literature. It is important to opt for a reliable strategy to prevent or reduce asphaltene problems, including, reservoir impairments, well and pipeline plugging, and separation problems at surface separation facilities [141]. There are also some suggested preventive strategies or methods that mitigate the process of asphaltene deposition that can be practiced as a remedial action for reservoirs that are prone to asphaltene depositions [141]. However, such actions take enormous costs, especially regarding the asphaltene deposits cleaning and loss of productivity (Table 1.3) [141].

The economic consequences of asphaltene deposition are significant if only the work over cost of well (due to the presence of H_2S) is considered, which costs up to $250,000 [9]. In a reservoir in Venezuela, asphaltene sludge deposited and caused partial or complete plugging of the well right after a temporary shut-in well for the acid stimulation treatment [142]. In the case of the Hassi Messaoud field, Algeria, the formation of asphaltene deposits in the tubing system was a very severe production issue that required tubing washing and scraping processes [143]. The problems associated with

Table 1.3 Typical additional costs due to asphaltene deposition problems [141].

Operation	Cost
Deposit removal	$300,000–$3,500,000 per well
Side track	$50,000,000 per well
Downtime	$700,000 per day (for a production of 7000 BPD)

asphaltene depositions annually cost billions of dollars, accounting for the reduced oil production, shuts in wells, incompetent production based on the reservoir capacity, and sedimentation and treatment processes [144]. In the Gulf of Mexico field, the rig intervention costs were estimated to be equal to $70 million per well because of asphaltene deposition and well shut-in. It was also estimated that this number could even increase to $100 million per well if the deposits are formed at the surface-controlled subsurface safety valve. For the case of 10,000 bbl/day production and an oil price of $50 per barrel, the cost of downtime would be around $500,000 per day [38], and the replacement of a lost well with a sidetrack will also approximately increase the cost up to $150 million [38]. The expensive cleaning and treatment processes are widely applied to prevent and remediate these deposits. Chemical treatments are among the techniques used in the Middle East where the injection of chemical additives annually costs about $31,000–46,000 per well; similarly, the treatment cost in a typical well in the Gulf of Mexico annually is about $330,000–390,000 per well. The imposition of these additional costs signals the need for the development of cost-effective approaches to control asphaltene deposition and ensure the high efficiency of the production process. In the Deep Horizon reservoirs, the production history and economics are widely affected by the asphaltene deposition. Even if asphaltene deposition does not occur in the primary and secondary stages, asphaltene can still deposit in tertiary recovery processes such as CO_2 injection, adding extra costs to the production [145]. The costs of steps/items such as the downtime, removal of asphaltene, and maintenance are considerable in heavy equipment and production or processing from a reservoir subjected to asphaltene; these parameters substantially impact the economy and recovery of a reservoir [146]. Considering the trend of oil production industry towards heavier and more asphaltenic reservoirs as well as the increase of implementing EOR methods' application such as the miscible gas injection, the influence of asphaltene control from the economic aspects becomes more crucial.

1.7 Research and development in asphaltenes

Asphaltene molecules have the potency to flocculate, precipitate, deposit, and finally, cease the continuity of the production process [8]. Recently, smart and novel strategies have been developed to overcome and prevent the asphaltene problems. Two perspectives are proposed for describing the asphaltene phase behavior. In the first theory, asphaltenes are considered as colloidal particles in crude oil and behave as particles in the colloidal system [141]. The second theory attributes the asphaltenes' behavior to their thermodynamic properties, revealing that asphaltenes are soluble in crudes [141].

Generally, less severe asphaltene deposition problems are reported for the production from the saturated reservoirs [141], which is related to more desolvation power in heavier crudes with higher asphaltene percentage than that in the crude oil containing lower asphaltene content. Based on the crude oil characteristics, three main strategies are recommended to tackle the asphaltene precipitation problems [141]:

- *Thermodynamic control* refers to maintaining the system's pressure and temperature away from the asphaltene phase envelope, where the solid particles are stable.
- *Kinetic control* is to control the kinetic parameters of solid particles in such a way that the precipitation of asphaltene particles slows down.
- *Mechanical control* refers to the periodical removal of the asphaltenes after their deposition.

One of the challenges in preventing asphaltene precipitation is to establish an appropriate wellbore condition with an efficient well monitoring system to monitor (and fix) any abnormalities in the system. A common strategy in production and flow assistance is the periodical stimulation of the wells to enhance the productivity index (PI), that is, the production rate over pressure drop. This type of treatment usually includes acid for stimulation, and xylene and mutual solvent for opening new paths and dissolving asphaltenes, respectively [141]. Mechanical, thermal, and chemical methods have been emerged and evolved to mitigate the asphaltene deposition. In the mechanical methods, different mechanical tools such as smooth-surface tubular, tubular brushes and scrapers, soluble and insoluble pigs and balls, free-floating pistons, and hollow rods are used to remove the asphaltenes in a pipe. In the thermal methods, asphaltenes are removed from tubular using bottom-hole heaters, microwaves, hot oil circulation, steam treatments, and heat liberating chemicals [141]. In the chemical methods, asphaltene inhibitors and solvents, such as aromatic hydrocarbons (e.g., xylene or toluene) are commonly used as asphaltene solvents. The main drawback of all the mentioned strategies is the lack of durability, and shortly after treatment the asphaltene particles agglomerate and deposit again. Another problem is that asphaltene deposition increases with pressure drop; this requires more treatment repetitions and, subsequently, it results in imposing extra costs and adverse effects on the PI of wells. Current environmental constraints and human safety regulations restrict the application of ongoing highly aromatic solvents, such as toluene and xylene, due to their low flash points, high toxicity, and low biodegradability.

1.8 Current and future status of asphaltenes

The characterization and behavior of crude oils have always been an area of concern in petroleum engineering. Peculiarly, the analysis and management of asphaltene particles are among the most problematic issues as the asphaltene precipitation/deposition causes numerous challenges in production and imposes an additional cost. Thus, the characterization of asphaltenes is a crucial step in the development of cost-effective strategies for the prediction and remediation of asphaltene problems. Due to the high complexity of the crude oil, as a hydrocarbon mixture, the determination of asphaltene properties, as individual components, becomes unfeasible. Numerous methods are developed to group several components of crude oil according to their boiling points, chemical structure, and polarity. The developed models have limitations, such as being time consuming, economically expensive, and analytically inaccurate. Asphaltenes are highly polarizable polyaromatic molecules, containing heteroatoms and metals, and can create nanoclusters that cause complexities and challenges concerning the asphaltenic crude oils [38]. Therefore, it is required to devise efficient techniques for analyzing asphaltene behavior and developing chemicals to prevent asphaltene aggregation and deposition. Several experiments are required to obtain asphaltene physiochemistry properties. Various techniques are designed to characterize the asphaltene chemical structure, such as fourier transform infrared (FTIR), Raman, and nuclear magnetic resonance (NMR) spectroscopy. Some methods are developed for analyzing the compositional elements, discovering thermal decomposition patterns, and calculating their molecular weight distribution. The onset point of asphaltene precipitation at the reservoir condition can be measured with a gravimetric method, acoustic resonance technology (ART), light scattering technique (LST), high-pressure microscopy (HPM), HPHT filtration, and quartz crystal resonance (QCR) method [38]. The LST and the HPM methods are the industry standards to measure the onset point of asphaltene precipitation. Most of the research studies are conducted using dead oil at ambient pressure and temperature due to the high costs of experiments in abnormal pressure and temperature conditions. Therefore, thermodynamic models can be alternatively used to evaluate the asphaltene precipitation behavior at elevated pressures and temperatures. The indirect methods are superior to direct techniques in different aspects such as applicability to the crude oils with either high or low asphaltene content, and the high sensitivity that enables detecting precipitated particles as small as 100 nm. The indirect techniques are also capable of quantifying the amount of precipitated asphaltene [147].

1.9 Final remarks

This chapter answers the general questions about asphaltene such as what is the asphaltene, what the asphaltene structure looks like, why the asphaltene precipitates and deposits, and what consequences it might have. It is mentioned that the asphaltene deposition results in oil production difficulty and imposes additional costs to the industry due to the production interruption and treatment process. The main theory of asphaltene aggregation such as the micellization theory is briefly described and the modelling methods of asphaltene precipitation and deposition are reviewed. The modelling methods are very helpful in the evaluation of asphaltene behaviors in the situations that are demanding and/or expensive to be tested in laboratory.

References

[1] O.C. Mullins, The asphaltenes, Annu. Rev. Anal. Chem. 4 (2011) 393–418.
[2] J.J. Adams, Asphaltene adsorption, a literature review, Energy Fuel 28 (5) (2014) 2831–2856.
[3] S. Subramanian, S. Simon, J. Sjöblom, Asphaltene precipitation models: a review, J. Dispers. Sci. Technol. 37 (7) (2016) 1027–1049.
[4] I. Wiehe, K. Liang, Asphaltenes, resins, and other petroleum macromolecules, Fluid Phase Equilib. 117 (1–2) (1996) 201–210.
[5] P. Luo, X. Wang, Y. Gu, Characterization of asphaltenes precipitated with three light alkanes under different experimental conditions, Fluid Phase Equilib. 291 (2) (2010) 103–110.
[6] S. Chopra, L. Lines, Introduction to this special section: heavy oil, Lead. Edge 27 (9) (2008) 1104–1106.
[7] M.S. Rana, et al., A review of recent advances on process technologies for upgrading of heavy oils and residua, Fuel 86 (9) (2007) 1216–1231.
[8] S.L. Kokal, S.G. Sayegh, Asphaltenes: the cholesterol of petroleum, in: Middle East Oil Show, Society of Petroleum Engineers, 1995.
[9] K.J. Leontaritis, G.A. Mansoori, Asphaltene deposition: a survey of field experiences and research approaches, J. Petrol. Sci. Eng. 1 (3) (1988) 229–239.
[10] J. Buckley, Y. Liu, Some mechanisms of crude oil/brine/solid interactions, J. Petrol. Sci. Eng. 20 (3–4) (1998) 155–160.

[11] C. Drummond, J. Israelachvili, Fundamental studies of crude oil–surface water interactions and its relationship to reservoir wettability, J. Petrol. Sci. Eng. 45 (1–2) (2004) 61–81.
[12] J.L. Mendoza de la Cruz, F.J. Arguelles-Vivas, V. Matias-Perez, C.D.L.A. Durán-Valencia, S. Lopez-Ramirez, Asphaltene-induced precipitation and deposition during pressure depletion on a porous medium: an experimental investigation and modeling approach, Energy Fuel 23 (11) (2009) 5611–5625.
[13] R. Syunyaev, et al., Adsorption of petroleum asphaltenes onto reservoir rock sands studied by near-infrared (NIR) spectroscopy, Energy Fuel 23 (3) (2009) 1230–1236.
[14] M. Vafaie-Sefti, S. Mousavi-Dehghani, Application of association theory to the prediction of asphaltene deposition: deposition due to natural depletion and miscible gas injection processes in petroleum reservoirs, Fluid Phase Equilib. 247 (1–2) (2006) 182–189.
[15] I. Gawel, D. Bociarska, P. Biskupski, Effect of asphaltenes on hydroprocessing of heavy oils and residua, Appl. Catal. Gen. 295 (1) (2005) 89–94.
[16] K. Akbarzadeh, et al., Asphaltenes—problematic but rich in potential, Oilfield Rev. 19 (2) (2007) 22–43.
[17] A. Cosultchi, P. Rossbach, I. Hernández-Calderon, XPS analysis of petroleum well tubing adherence, Surf. Interface Anal. 35 (3) (2003) 239–245.
[18] C.H. Bartholomew, Mechanisms of catalyst deactivation, Appl. Catal. Gen. 212 (1–2) (2001) 17–60.
[19] T. Masuda, K. Hashimoto, Deactivation of zeolite catalysts by coke, in: Deactivation and Testing of Hydrocarbon-Processing Catalysts, vol. 634, American Chemical Society, 1996, pp. 62–76.
[20] J. Ancheyta, et al., Extraction and characterization of asphaltenes from different crude oils and solvents, Energy Fuel 16 (5) (2002) 1121–1127.
[21] M. Absi-Halabi, A. Stanislaus, D. Trimm, Coke formation on catalysts during the hydroprocessing of heavy oils, Appl. Catal. 72 (2) (1991) 193–215.
[22] F.M. Faus, P. Grange, B. Delmon, Influence of asphaltene deposition on catalytic activity of cobalt molybdenum on alumina catalysts, Appl. Catal. 11 (2) (1984) 281–293.
[23] C. Marchal, et al., Asphaltene diffusion and adsorption in modified NiMo alumina catalysts followed by ultraviolet (UV) spectroscopy, Energy Fuel 24 (8) (2010) 4290–4300.
[24] M. Barcenas, et al., Study of medium effect on asphaltene agglomeration inhibitor efficiency, Energy Fuel 22 (3) (2008) 1917–1922.
[25] D.M. Bilden, V.E. Jones, Asphaltene Adsorption Inhibition Treatment, 2000. Google Patents US6051535A United States.
[26] C.-L. Chang, H. Scott Fogler, Peptization and coagulation of asphaltenes in apolar media using oil-soluble polymers, Fuel Sci. Technol. Int. 14 (1–2) (1996) 75–100.
[27] G. Gonzalez, A. Middea, Peptization of asphaltene by various oil soluble amphiphiles, Colloids Surf. 52 (1991) 207–217.
[28] O. León, et al., Study of the adsorption of alkyl benzene-derived amphiphiles on asphaltene particles, Langmuir 15 (22) (1999) 7653–7657.
[29] J.-A. Östlund, et al., Characterization of fractionated asphaltenes by UV–vis and NMR self-diffusion spectroscopy, J. Colloid Interface Sci. 271 (2) (2004) 372–380.
[30] J. Argillier, et al., Heavy oil dilution, in: SPE International Thermal Operations and Heavy Oil Symposium, Society of Petroleum Engineers, 2005.
[31] S.W. Hasan, M.T. Ghannam, N. Esmail, Heavy crude oil viscosity reduction and rheology for pipeline transportation, Fuel 89 (5) (2010) 1095–1100.
[32] C.K. Yeung, System for the Decontamination of Asphaltic Heavy Oil and Bitumen, 2009. Google Patents, US7625466B2, United States.

[33] U.G. Romanova, et al., Investigation of oil sands froth treatment, Can. J. Chem. Eng. 82 (4) (2004) 710–721.
[34] Y. Zhao, F. Wei, Simultaneous removal of asphaltenes and water from water-in-bitumen emulsion: I. Fundamental development, Fuel Process. Technol. 89 (10) (2008) 933–940.
[35] A. Duong, K.J. Smith, A model of ceramic membrane fouling during heavy oil ultrafiltration, Can. J. Chem. Eng. 75 (6) (1997) 1122–1129.
[36] H.-L. Wang, et al., Upgrading residue by carbon rejection in a fluidized-bed reactor and its multiple lump kinetic model, Ind. Eng. Chem. Res. 50 (22) (2011) 12501–12511.
[37] J.P. Rankin, et al., Methods for Upgrading of Contaminated Hydrocarbon Streams, 2012. Google Patents, US8197671B2, United States.
[38] F.M. Vargas, M. Tavakkoli, Asphaltene Deposition: Fundamentals, Prediction, Prevention, and Remediation, CRC Press, 2018.
[39] J.B. Boussingault, Mémoire sur l'influence des défrichements dans la diminution des cours d'eau, Ann. Chim. 64 (1837) 113–141.
[40] O.C. Mullins, et al., Asphaltenes, Heavy Oils, and Petroleomics, Springer Science AMPamp; Business Media, 2007.
[41] K. Van Nes, H.A. Van Westen, Aspects of the Constitution of Mineral Oils, Elsevier Pub. Co, 1951.
[42] T.F. Yen, G.V. Chilingarian, Asphaltenes and Asphalts, 2, Elsevier, 2000.
[43] C. Zhang, J. Huang, Drilling and Gas Recovery Technology in Ancient China, Zigong Salt Industry Museum, Zigong, 1997, p. 44.
[44] H. Abraham, Asphalts and Allied Substances: Industrial Raw Materials, vol. 2, Van Nostrand, 1961.
[45] T. Yen, The realms and definitions of asphaltenes, in: Developments in Petroleum Science, Elsevier, 2000, pp. 7–28.
[46] H. Groenzin, O.C. Mullins, Asphaltene molecular size and structure, J. Phys. Chem. A 103 (50) (1999) 11237–11245.
[47] O.C. Mullins, The modified Yen model, Energy Fuel 24 (4) (2010) 2179–2207.
[48] J. Murgich, J.A. Abanero, O.P. Strausz, Molecular recognition in aggregates formed by asphaltene and resin molecules from the Athabasca oil sand, Energy Fuel 13 (2) (1999) 278–286.
[49] J.M. Sheremata, et al., Quantitative molecular representation and sequential optimization of athabasca asphaltenes, Energy Fuel 18 (5) (2004) 1377–1384.
[50] A.H. Alshareef, et al., Formation of archipelago structures during thermal cracking implicates a chemical mechanism for the formation of petroleum asphaltenes, Energy Fuel 25 (5) (2011) 2130–2136.
[51] A. Karimi, et al., Quantitative evidence for bridged structures in Asphaltenes by thin film pyrolysis, Energy Fuel 25 (8) (2011) 3581–3589.
[52] H. Sabbah, et al., Evidence for island structures as the dominant architecture of Asphaltenes, Energy Fuel 25 (4) (2011) 1597–1604.
[53] B. Schuler, et al., Unraveling the molecular structures of Asphaltenes by atomic force microscopy, J. Am. Chem. Soc. 137 (31) (2015) 9870–9876.
[54] B.A. Watson, M.A. Barteau, Imaging of petroleum asphaltenes using scanning tunneling microscopy, Ind. Eng. Chem. Res. 33 (10) (1994) 2358–2363.
[55] G. Zajac, N. Sethi, J. Joseph, Molecular imaging of petroleum asphaltenes by scanning-tunneling-microscopy-verification of structure from C-13 and proton nuclear-magnetic-resonance data, Scanning Microsc. 8 (3) (1994) 463–470.
[56] L. Gross, et al., The chemical structure of a molecule resolved by atomic force microscopy, Science 325 (5944) (2009) 1110–1114.

[57] L. Gross, et al., Bond-order discrimination by atomic force microscopy, Science 337 (6100) (2012) 1326–1329.
[58] L. Gross, et al., Organic structure determination using atomic-resolution scanning probe microscopy, Nat. Chem. 2 (10) (2010) 821–825.
[59] K.Ø. Hanssen, et al., A combined atomic force microscopy and computational approach for the structural elucidation of breitfussin A and B: highly modified halogenated dipeptides from *Thuiaria breitfussi*, Angew. Chem. 124 (49) (2012) 12404–12407.
[60] J. Van Der Lit, et al., Suppression of electron–vibron coupling in graphene nanoribbons contacted via a single atom, Nat. Commun. 4 (1) (2013) 1–6.
[61] B. Schuler, et al., From perylene to a 22-ring aromatic hydrocarbon in one-pot, Angew. Chem. 126 (34) (2014) 9150–9152.
[62] D.G. de Oteyza, et al., Direct imaging of covalent bond structure in single-molecule chemical reactions, Science 340 (6139) (2013) 1434–1437.
[63] A. Riss, et al., Local electronic and chemical structure of oligo-acetylene derivatives formed through radical cyclizations at a surface, Nano Lett. 14 (5) (2014) 2251–2255.
[64] L. Gross, et al., High-resolution molecular orbital imaging using a p-wave STM tip, Phys. Rev. Lett. 107 (8) (2011), 086101.
[65] J. Repp, et al., Molecules on insulating films: scanning-tunneling microscopy imaging of individual molecular orbitals, Phys. Rev. Lett. 94 (2) (2005), 026803.
[66] G.K. Wong, T.F. Yen, An electron spin resonance probe method for the understanding of petroleum asphaltene macrostructure, J. Petrol. Sci. Eng. 28 (1) (2000) 55–64.
[67] O. Leon, et al., Asphaltenes: structural characterization, self-association, and stability behavior, Energy Fuel 14 (1) (2000) 6–10.
[68] N.B. Joshi, O.C. Mullins, A. Jamaluddin, J. Creek, J. McFadden, Asphaltene precipitation from live crude oil, Energy Fuel 15 (4) (2001) 976–998.
[69] J.W. Shirokoff, M.N. Siddiqui, M.F. Ali, Characterization of the structure of Saudi crude asphaltenes by X-ray diffraction, Energy Fuel 11 (3) (1997) 561–565.
[70] A. Sharma, et al., Probing order in asphaltenes and aromatic ring systems by HRTEM, Energy Fuel 16 (2) (2002) 490–496.
[71] E. Buenrostro-Gonzalez, et al., The overriding chemical principles that define asphaltenes, Energy Fuel 15 (4) (2001) 972–978.
[72] H. Alboudwarej, et al., Sensitivity of asphaltene properties to separation techniques, Energy Fuel 16 (2) (2002) 462–469.
[73] L.B. Gutiérrez, et al., Fractionation of asphaltene by complex formation with p-nitrophenol. A method for structural studies and stability of asphaltene colloids, Energy Fuel 15 (3) (2001) 624–628.
[74] H.W. Yarranton, H. Alboudwarej, R. Jakher, Investigation of asphaltene association with vapor pressure osmometry and interfacial tension measurements, Ind. Eng. Chem. Res. 39 (8) (2000) 2916–2924.
[75] M. Guiliano, et al., Supercritical fluid extraction of bal 150 crude oil asphaltenes, Energy Fuel 14 (1) (2000) 89–94.
[76] H. Groenzin, O.C. Mullins, Molecular size and structure of asphaltenes from various sources, Energy Fuel 14 (3) (2000) 677–684.
[77] B. Fuhr, L. Holloway, A. Hammami, Analytical considerations related to asphaltenes and waxes in the same crudes, Energy Fuel 13 (2) (1999) 336–339.
[78] M. Domin, et al., A comparative study of bitumen molecular-weight distributions, Energy Fuel 13 (3) (1999) 552–557.
[79] L. Carbognani, M. Orea, M. Fonseca, Complex nature of separated solid phases from crude oils, Energy Fuel 13 (2) (1999) 351–358.

[80] L. Artok, et al., Structure and reactivity of petroleum-derived asphaltene, Energy Fuel 13 (2) (1999) 287–296.
[81] S. Peramanu, B.B. Pruden, P. Rahimi, Molecular weight and specific gravity distributions for Athabasca and cold lake bitumens and their saturate, aromatic, resin, and asphaltene fractions, Ind. Eng. Chem. Res. 38 (8) (1999) 3121–3130.
[82] M. Tojima, et al., Effect of heavy asphaltene on stability of residual oil, Catal. Today 43 (3–4) (1998) 347–351.
[83] J. Miller, et al., Subfractionation and characterization of Mayan asphaltene, Energy Fuel 12 (6) (1998) 1290–1298.
[84] J.D. McLean, P.K. Kilpatrick, Comparison of precipitation and extrography in the fractionation of crude oil residua, Energy Fuel 11 (3) (1997) 570–585.
[85] S.I. Andersen, Separation of asphaltenes by polarity using liquid-liquid extraction, Pet. Sci. Technol. 15 (1–2) (1997) 185–198.
[86] S.I. Andersen, Effect of precipitation temperature on the composition of n-heptane asphaltenes part 2, Fuel Sci. Technol. Int. 13 (5) (1995) 579–604.
[87] B. Fuhr, et al., Properties of asphaltenes from a waxy crude, Fuel 70 (11) (1991) 1293–1297.
[88] R.S.F.I. Adarme, M.M. Jonson, D.R. Kidd, Phillips, J.E.S.o.R.U.A. M. D, D. Shaw, Chemical Society: Washington, 1990, pp. 614–618.
[89] O.C. Mullins, E.Y. Sheu, Structures and Dynamics of Asphaltenes, Springer US, 2013.
[90] O.L. Chapman, Spectrometric identification of organic compounds, J. Am. Chem. Soc. 85 (20) (1963) 3316.
[91] A.T. Castro, NMR and FTIR characterization of petroleum residues: structural parameters and correlations, J. Braz. Chem. Soc. 17 (2006) 1181–1185.
[92] A. Yen, Y.R. Yin, S. Asomaning, Evaluating asphaltene inhibitors: laboratory tests and field studies, in: SPE International Symposium on Oilfield Chemistry, Society of Petroleum Engineers, Houston, Texas, 2001, p. 7.
[93] A. Hirschberg, et al., Influence of temperature and pressure on Asphaltene flocculation, Soc. Pet. Eng. J. 24 (03) (1984) 283–293.
[94] S. Zendehboudi, et al., Asphaltene precipitation and deposition in oil reservoirs – technical aspects, experimental and hybrid neural network predictive tools, Chem. Eng. Res. Des. 92 (5) (2014) 857–875.
[95] J.H. Pacheco-Sanchez, G. Ali Mansoori, Prediction of the phase behavior of asphaltene micelle/aromatic hydrocarbon systems, Pet. Sci. Technol. 16 (3–4) (1998) 377–394.
[96] S. Mousavi-Dehghani, et al., An analysis of methods for determination of onsets of asphaltene phase separations, J. Petrol. Sci. Eng. 42 (2–4) (2004) 145–156.
[97] P.J. Missel, et al., Thermodynamic analysis of the growth of sodium dodecyl sulfate micelles, J. Phys. Chem. 84 (9) (1980) 1044–1057.
[98] K.L. Mittal, Micellization, Solubilization, and Microemulsions, vol. 2, Springer Science AMPamp; Business Media, 2012.
[99] O.C. Mullins, et al., Advances in Asphaltene science and the Yen–Mullins model, Energy Fuel 26 (7) (2012) 3986–4003.
[100] J.P. Pfeiffer, R.N.J. Saal, Asphaltic bitumen as colloid system, J. Phys. Chem. 44 (2) (1940) 139–149.
[101] S.L. Kokal, S.G. Sayegh, Asphaltenes: the cholesterol of petroleum, in: Middle East Oil Show, Society of Petroleum Engineers, Bahrain, 1995, p. 13.
[102] W. Chaisoontornyotin, A.W. Bingham, M.P. Hoepfner, Reversibility of Asphaltene precipitation using temperature-induced aggregation, Energy Fuel 31 (4) (2017) 3392–3398.

[103] F.M. Vargas, et al., On the development of an enhanced method to predict asphaltene precipitation, in: Offshore Technology Conference, Offshore Technology Conference, Houston, Texas, 2014, p. 19.
[104] C. Zheng, et al., Dissolution and suspension of asphaltenes with ionic liquids, Fuel 238 (2019) 129–138.
[105] E.F. Ghloum, M. Al-Qahtani, A. Al-Rashid, Effect of inhibitors on asphaltene precipitation for Marrat Kuwaiti reservoirs, J. Petrol. Sci. Eng. 70 (1) (2010) 99–106.
[106] H. Alrashidi, S. Afra, H.A. Nasr-El-Din, Application of natural fatty acids as asphaltenes solvents with inhibition and dispersion effects: a mechanistic study, J. Petrol. Sci. Eng. 172 (2019) 724–730.
[107] R. Tanaka, et al., Characterization of Asphaltene aggregates using X-ray diffraction and small-angle X-ray scattering, Energy Fuel 18 (4) (2004) 1118–1125.
[108] S. Wang, et al., Interaction forces between Asphaltene surfaces in organic solvents, Langmuir 26 (1) (2010) 183–190.
[109] K. Karan, et al., Evaluation of asphaltene instability and a chemical control during production of live oils, Pet. Sci. Technol. 21 (3–4) (2003) 629–645.
[110] K. Indo, et al., Asphaltene nanoaggregates measured in a live crude oil by centrifugation, Energy Fuel 23 (9) (2009) 4460–4469.
[111] M.R. Gray, et al., Supramolecular assembly model for aggregation of petroleum asphaltenes, Energy Fuel 25 (7) (2011) 3125–3134.
[112] R. De Boer, et al., Screening of crude oils for asphalt precipitation: theory, practice, and the selection of inhibitors, SPE Prod. Facil. 10 (01) (1995) 55–61.
[113] K.J. Leontaritis, G.A. Mansoori, Asphaltene flocculation during oil production and processing: a thermodynamic collodial model, in: SPE International Symposium on Oilfield Chemistry, Society of Petroleum Engineers, 1987.
[114] D. Powers, et al., Regular solution based approach to modeling asphaltene precipitation from native and reacted oils: part 1, molecular weight, density, and solubility parameter distributions of asphaltenes, Fuel 178 (2016) 218–233.
[115] H. Alboudwarej, et al., Regular solution model for asphaltene precipitation from bitumens and solvents, AICHE J. 49 (11) (2003) 2948–2956.
[116] L. Nghiem, et al., Efficient modelling of asphaltene precipitation, in: SPE Annual Technical Conference and Exhibition, Society of Petroleum Engineers, Houston, Texas, 1993. 3-6 October.
[117] L.X. Nghiem, D.A. Coombe, S. Ali, Compositional simulation of asphaltene deposition and plugging, in: SPE Annual Technical Conference and Exhibition, Society of Petroleum Engineers, New Orleans, Louisiana, 1998. 27-30 September.
[118] P.J. Flory, Thermodynamics of high polymer solutions, J. Chem. Phys. 10 (1) (1942) 51–61.
[119] N.E. Burke, R.E. Hobbs, S.F. Kashou, Measurement and modeling of Asphaltene precipitation (includes associated paper 23831), J. Petrol. Tech. 42 (11) (1990) 1,440–1,446.
[120] S.L. Kokal, et al., Measurement and correlation of asphaltene precipitation from heavy oils by gas injection, J. Can. Pet. Technol. (1992) **31**(04).
[121] Z. Novosad, T. Costain, Experimental and modeling studies of asphaltene equilibria for a reservoir under CO_2 injection, in: SPE Annual Technical Conference and Exhibition, Society of Petroleum Engineers, New Orleans, Louisiana, 1990. 23-26 September.
[122] F. Chung, P. Sarathi, R. Jones, Modeling of Asphaltene and Wax Precipitation, National Inst. for Petroleum and Energy Research, Bartlesville, OK (USA), 1991.
[123] A.I. Victorov, A. Firoozabadi, Thermodynamic micellizatin model of asphaltene precipitation from petroleum fluids, AICHE J. 42 (6) (1996) 1753–1764.

[124] H. Rassamdana, et al., Asphalt flocculation and deposition: I. the onset of precipitation, AICHE J. 42 (1) (1996) 10–22.
[125] Y.-F. Hu, et al., A study on the application of scaling equation for asphaltene precipitation, Fluid Phase Equilib. 171 (1–2) (2000) 181–195.
[126] A.K. Menshad, et al., Developing of scaling equation with function of pressure to determine onset of asphaltene precipitation, J. Jpn. Pet. Inst. 51 (2) (2008) 102–106.
[127] M. Bagheri, et al., Developing a new scaling equation for modelling of asphaltene precipitation, in: Canadian International Petroleum Conference, Petroleum Society of Canada, Calgary, Alberta, 2009. 16-18 June.
[128] B.S. Soulgani, et al., A new thermodynamic scale equation for modelling of asphaltene precipitation form live oil, in: Canadian International Petroleum Conference, Petroleum Society of Canada, Calgary, Alberta, 2009. 16-18 June.
[129] G. Zahedi, et al., Prediction of asphaltene precipitation in crude oil, J. Petrol. Sci. Eng. 68 (3–4) (2009) 218–222.
[130] M.A. Ahmadi, Prediction of asphaltene precipitation using artificial neural network optimized by imperialist competitive algorithm, J. Pet. Explor. Prod. Technol. 1 (2–4) (2011) 99–106.
[131] S. Alimohammadi, J.S. Amin, E. Nikooee, Estimation of asphaltene precipitation in light, medium and heavy oils: experimental study and neural network modeling, Neural Comput. Applic. 28 (4) (2017) 679–694.
[132] S.M. Hashmi, A. Firoozabadi, Effect of dispersant on asphaltene suspension dynamics: aggregation and sedimentation, J. Phys. Chem. B 114 (48) (2010) 15780–15788.
[133] R. Miri, et al., Experimental and numerical modeling study of gravity drainage considering asphaltene deposition, Ind. Eng. Chem. Res. 53 (28) (2014) 11512–11526.
[134] S. Moradi, et al., Experimental investigation and modeling of asphaltene precipitation due to gas injection, Iran. J. Chem. Chem. Eng. 31 (1) (2012) 89–98.
[135] J. Sayyad Amin, et al., Prediction of asphaltene precipitation: learning from data at different conditions, Energy Fuel 24 (7) (2010) 4046–4053.
[136] S. Collins, J. Melrose, Adsorption of asphaltenes and water on reservoir rock minerals, in: SPE Oilfield and Geothermal Chemistry Symposium, Society of Petroleum Engineers, 1983.
[137] S. Dubey, M. Waxman, Asphaltene adsorption and desorption from mineral surfaces, SPE Reserv. Eng. 6 (03) (1991) 389–395.
[138] G. Gonzalez, A.M. Travalloni-Louvisse, Adsorption of asphaltenes and its effect on oil production, SPE Prod. Facil. 8 (02) (1993) 91–96.
[139] C. Gruesbeck, R. Collins, Entrainment and deposition of fine particles in porous media, Soc. Pet. Eng. J. 22 (06) (1982) 847–856.
[140] L.M. Cenegy, Survey of successful world-wide asphaltene inhibitor treatments in oil production fields, in: SPE Annual Technical Conference and Exhibition, Society of Petroleum Engineers, New Orleans, Louisiana, 2001, p. 7. 30 September-3 October.
[141] A. Al-Qasim, et al., Asphaltene precipitation: a review on remediation techniques and prevention strategies, in: Abu Dhabi International Petroleum Exhibition AMPamp; Conference, Society of Petroleum Engineers, Abu Dhabi, UAE, 2018, p. 9. 12–15 November, Abu Dhabi, UAE.
[142] P.M. Lichaa, L. Herrera, Electrical and other effects related to the formation and prevention of asphaltene deposition problem in venezuelan crudes, in: SPE Oilfield Chemistry Symposium, Society of Petroleum Engineers, Dallas, Texas, 1975, p. 19.
[143] C.E. Haskett, M. Tartera, A practical solution to the problem of Asphaltene deposits-Hassi Messaoud field, Algeria, J. Petrol. Tech. 17 (04) (1965) 387–391.
[144] A.M. Atta, A.M. Elsaeed, Use of rosin-based nonionic surfactants as petroleum crude oil sludge dispersants, J. Appl. Polym. Sci. 122 (1) (2011) 183–192.

[145] R.N. Tuttle, High-pour-point and asphaltic crude oils and condensates, J. Petrol. Tech. 35 (06) (1983) 1,192–1,196.
[146] G.A. Mansoori, Asphaltene deposition: an economic challenge in heavy petroleum crude utilization and processing, OPEC Rev. 12 (1) (1988) 103–113.
[147] M. Tavakkoli, et al., Indirect method: a novel technique for experimental determination of asphaltene precipitation, Energy Fuel 29 (5) (2015) 2890–2900.

CHAPTER TWO

Asphaltene precipitation/ deposition control methods

2.1 Introduction

Precipitation and deposition of asphaltene occur as a result of variations in operating conditions, or composition of the crude oil during the reservoir depletion and EOR/IOR processes [1, 2]. The asphaltene-related problems can be handled before or after its precipitation and deposition. Fig. 2.1 shows an overview of the asphaltene inhibition, mitigation, and removal methods. Early solutions for the asphaltene deposition problem focused on asphaltene removal [3, 4]. Conventional methods include employing scrapers and aromatic solvents which are time consuming, costly, and eco-destructive [4]. Alternatively, physical and chemical techniques are proposed to inhibit or mitigate the asphaltene deposition; methods such as ultrasonication and injection of chemicals are used to reduce the asphaltene problems. The effectiveness of inhibition, mitigation, and removal methods is related to the properties of the crude oil and its asphaltene, thermodynamic conditions, compaction and hardness of deposition, and influential parameters of treatment methods [5]. In this chapter, conventional physical and chemical methods for asphaltene treatment are introduced, and their underlying inhibitory/removal mechanisms are discussed (Fig. 2.1).

2.2 Preventing asphaltene precipitation: Description, methods, and mechanisms

Preventing asphaltene precipitation includes a process that restricts the precipitation, aggregation, and finally deposition of the asphaltene particles through physical or chemical methods. The physical methods include manipulating pressure/temperature or using more advanced methods such as ultrasonic waves to keep the asphaltene suspended without any chemical

Asphaltene Deposition Control by Chemical Inhibitors
https://doi.org/10.1016/B978-0-323-90510-7.00002-1

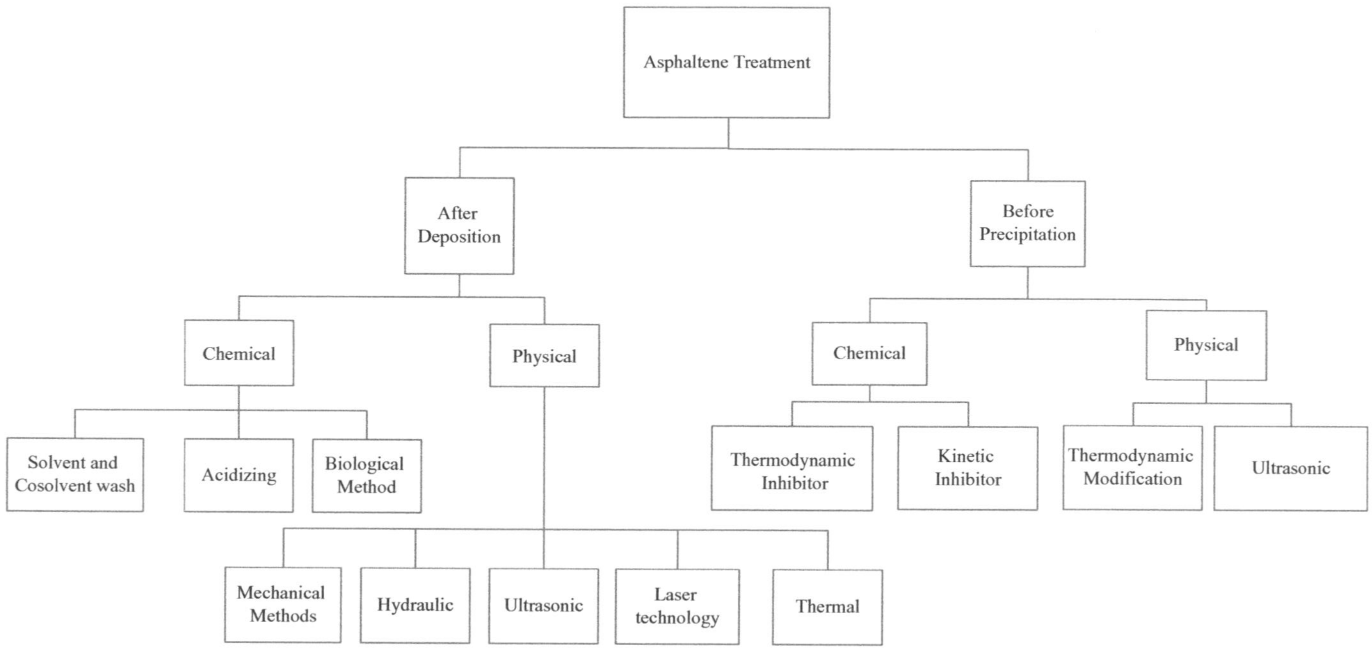

Fig. 2.1 Flowchart of asphaltene treatment.

reaction. However, the chemical methods change the chemistry or equilibrium conditions for the mixture by changing the composition or adding new components.

2.2.1 Chemical methods

Preventive (or inhibitive) chemical methods include the addition of a component/substance to the oil and changing chemical equilibrium. The chemical additive can be an extracted part of the crude oil (such as resin) or any other chemical substances (such as surfactants) that can attach to the asphaltenes and stabilize their suspension in oil media. The focus of this book is on chemical inhibitors; thus, different inhibitor types, inhibitory mechanisms, and experiments to assess the effectiveness of asphaltene inhibitors will be discussed. In classification, the chemical inhibitors can be categorized to thermodynamic inhibitors and kinetic inhibitors.

2.2.1.1 Thermodynamic inhibitors

Thermodynamic inhibitors change the oil composition (the ratio of saturation, aromatic, resin, and asphaltene). The resin and aromatic parts keep the asphaltene part dissolved/suspended in the oil while the asphaltene part is insoluble in the saturation part. The stability of a colloid system can be quantified by colloidal instability index (CII); for the colloid systems containing asphaltenes, the CII is defined by the ratio of saturate plus asphaltene mass over the summation of resin and aromatic mass. If $CII < 0.7$, the oil is stable while for $CII > 0.9$, the oil is unstable. Generally, by changing the oil composition, the phase envelope and the asphaltene envelope will change, and the thermodynamic conditions (pressure and temperature) of the target oil will move away from the unstable region of asphaltene envelope. Resins, deasphalted oils, vegetable oil, and some of the refinery products are used as the thermodynamic inhibitors. The addition of resins and deasphalted oil increases the portion of aromatic and resin parts, which changes the phase and asphaltene envelopes to inhibit asphaltene precipitation and deposition. The main disadvantage of the thermodynamic inhibitors is their short durability.

2.2.1.2 Kinetic inhibitors

Solvents with or without cosolvents were initially used to dissolve the asphaltene depositions. Over time, researchers realized that some of the cosolvents could stabilize the asphaltene suspension and inhibit its deposition. In the crude oil, resins stabilize the asphaltenes by adhering to the asphaltene particles and creating steric hindrance around the asphaltene. The effective cosolvents have a similar inhibitory mechanism as the resins. An asphaltene inhibitor should be able to interact with the asphaltene particles by intermolecular forces such as $\pi - \pi$, hydrogen bond, van der Waals, and polar bond while preventing the asphaltene particles from association (aggregation). The inhibitor efficiency is affected by the asphaltene and the dispersed medium characteristics. Although synthetic oils do not necessarily represent a realistic dispersed medium, they are used in the assessment of asphaltene inhibitor effectiveness. In general, an inhibitor might be effective with a specific oil and ineffective with another oil. In the screening of chemical additives, it is highly desirable to find chemicals that are effective with a broad range of crude oils [5]. The kinetic inhibitors can be divided into commercial and noncommercial ones. The commercial inhibitors are a mixture of different concentrations of inhibitors which are developed specifically for a crude oil. They are usually effective with the tested oil (or perhaps with a similar oil composition); due to confidentiality purposes, their characterizations are usually not reported in the literature [6]. The noncommercial inhibitors have chemical functional groups that are chosen based on the potential to form bonding with asphaltenes and keep them suspended. They can be divided into surfactants, polymers, and nanoparticles with subsequent subdivisions. Surfactants are divided into ionic and nonionic components; the ionic surfactants can be of the cationic, anionic, or amphoteric type. Nanoparticles are also divided into acidic, basic, and amphoteric.

2.2.2 Physical methods

The effectiveness of the chemical methods for asphaltene inhibition is uncertain and costly, which motivates researchers to inhibit asphaltene precipitation using physical methods. The physical methods are still under investigation and have not yet been proven successful compared to the chemical methods. The aims of physical methods are breaking, destructing, detaching, and maintaining the dispersivity of the asphaltene particles.

2.2.2.1 Pressure and temperature

Asphaltene precipitation is governed by the thermodynamics of the system; factors such as pressure, temperature, and composition dictate the reservoir fluid's phase behavior and production conditions [5]. If the oil thermodynamic condition falls inside the asphaltene envelope in the P-T diagram, asphaltene will precipitate and deposit (Fig. 2.2). There are production strategies to stabilize or minimize the asphaltene precipitation. Insulation and

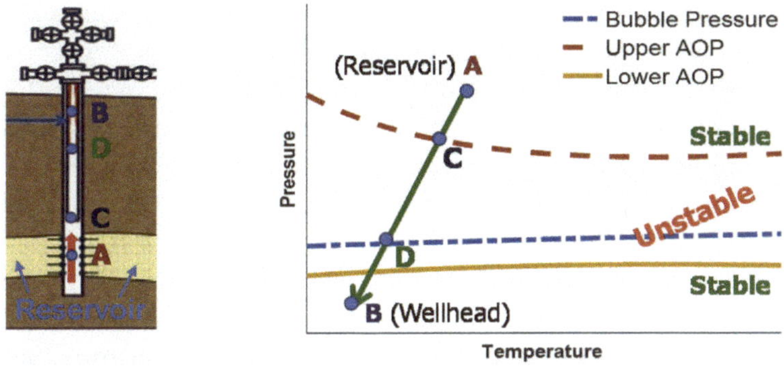

Fig. 2.2 Asphaltene onset pressure envelope for a particular oil in a pressure-temperature plot [7].

steam injection are the options that can change the temperature distribution between the bottom hole and wellhead condition. Moreover, changing the wellhead pressure alternates the pressure profile along the well or even near the wellbore. One strategy to avoid the deposition when the pressure falls bellow the upper asphaltene onset pressure curve (during the production) is decreasing the pressure suddenly, using venture (e.g., see Fig. 2.2). Additionally, the reservoir pressure can be maintained by injecting fluids like water or gas [5]. These techniques are intended to keep the pressure and temperature out of the asphaltene deposition/precipitation phase envelope. When manipulating the reservoir thermodynamic conditions, the economic and technical aspects of the reservoir should also be considered.

2.2.2.2 Ultrasonic

Ultrasonic waves have a variety of applications in the industry [8]. The effectiveness of ultrasonic waves as a promising physical method in the oil industry was realized accidentally by observing the changes in the level of water/ oil wells after seismic waves (and an earthquake) [8]. Ultrasonic waves can stretch the molecules' bonding while passing through a liquid. When the wave energy overcomes the intermolecular bonding energy, a void vacuum space is created, called "cavity." Because the pressure inside the cavities is lower than the fluid vapor pressure, the light components of crude oil are vaporized. Then, the gas bubbles grow due to the low compressibility of liquid, and finally, burst with releasing heat and energy. This process is called *Cavitation*, which is the main inhibitory mechanism of ultrasonic waves. This phenomenon not only decomposes asphaltene but also produces hydrocarbon radicals, which increase the probability of heavy oil cracking, both of which cause viscosity reduction [9–11]. The viscosity reduction using ultrasonic waves omits unnecessary energy consumption and cage effect in comparison with the thermal methods of viscosity reduction (pyrolysis) [9,10]. The ultrasonic wave frequency, power, and treatment time are considered as three important parameters of this method. These parameters should be optimized to make the treatment economically viable [12]. In a technical perspective, using longer excitation is more effective as it disperses the asphaltene to smaller particles [13,14]. Although substituting chemical methods with the ultrasonic method has both financial and operational advantages, it requires additional surface facilities such as electric supply that increases the operating and capital costs. Nevertheless, the energy consumption in the ultrasonic treatment process is about 43% lower than that in other vis-breaking processes [9].

2.3 Asphaltene removal: Description, methods, and mechanisms

Based on the thermodynamic conditions, the asphaltene can deposit in a reservoir, wellbore, production well, and surface process facilities. The asphaltene deposition increases pressure drop and decreases the production flow rate; in a severe case of blockage, it stops production. The treatment of deposited asphaltene is challenged by diagnosing the major deposition (blockage) location, for which the treatment might be different. The treatment methods are categorized by: chemical, physical, thermal, and biological. The chemical and physical methods are considered as common methods. Generally, the chemical methods employ aromatic solvents, and the physical methods use mechanical approaches. The thermal methods are more common to dissolve wax (rather than asphaltene) deposition, but there are a few reports of dissolving the asphaltene deposition using thermal methods. The biological methods are novel and still under investigations at the lab scale; although the biological methods are environmentally friendly, they have considerably slower treatment rates.

2.3.1 Chemical

The asphaltene is defined as a part of the oil that is soluble in aromatic solvents such as toluene and xylene. Therefore, aromatic solvents are considered as trivial solvents for the dissolution of asphaltene deposition. They are not environmentally friendly and impose hazardous impacts on the environment and human health. Therefore, novel eco-destructive chemicals such as biodegradable microorganisms are under investigation.

2.3.1.1 Solvent wash

Solvent wash is one of the most popular choices due to its applicability in the pipeline, well, and the wellbore facilities. This method does not need high-technology facilities and can be implemented using a simple coil tube injection system. The screening solvent type, injection rate, solvent volume, and soak time are determined in the lab [15]. The solubility of asphaltene in the solvent can be evaluated from solubility theory as described by Hansen. Chemicals with similar Hansen solubility parameters (HSP) form stable solutions. Reported values for the asphaltene Hansen solubility parameter are approximately 20 $MPa^{1/2}$; thus, a solvent with HSP close to 20 $MPa^{1/2}$ dissolve asphaltene [5]. Although toluene and xylene are efficient solvents, their health, safety, and environment hazards limit their large-scale applicability. The chemical method is costly and imposes additional operational costs due to corrosion and formation of water/oil emulsions [16]. Due to the high cost of conventional solvents, and their low durability and toxicity, some companies have abandoned them, substituting them with aromatic naphtha (commercially known as A150) [5]. In 2003, terpene-based environmentally friendly solvents were introduced for asphaltene removal. However, they were not as powerful as aromatic solvents [17]. Some researchers propose water-based solvents that feature higher efficiency in comparison with aromatic solvents [18]. In addition, the use of additives such as amines and acids is suggested to improve the asphaltene removal efficiency. All used chemicals dissolve the deposited asphaltenes in the formation and based on the type of asphaltene, the selection of a suitable chemical enhances the dissolution power [5]. Dispersants are other types of cosolvents that are used to improve solvent efficiency [16,19–21]. The dispersants do not dissolve the asphaltene and only inhibit their growth by adhering to the asphaltene solid interface [18].

2.3.1.2 Acidizing

Acidizing and acid washing are near wellbore treatment methods to cease asphaltene deposition. Acids do not dissolve the deposited asphaltene; rather, they dissolve the rock and create new paths (wormholes) for the oil flow. The efficiency of this method decreases by repetition since this method does not remove the asphaltene.

2.3.1.3 *Biological methods*

Different microorganisms, including anaerobic bacteria, aerobic bacteria, and fungus can decompose asphaltene mixtures; these should be studied and grown in the laboratory before using in the damaged area [22]. This method is still testing at the lab scale and is recognized as a slow process. It takes months to years for the microorganisms to consume the asphaltene deposits. Nevertheless, when this bioremediation process finds viability at the industrial scale, it can surpass other asphaltene removal methods because of its environmentally friendly and safe nature. Controlling the rate of microorganism's growth and effective reactions of consuming asphaltene are important in biological remediation methods. In screening the microorganisms, it is essential that microorganisms be resistant and efficient at the downhole condition [16,23].

2.3.2 Physical

Referring to disadvantages of the chemical asphaltene inhibitors, researchers have also focused on physical methods. The physical methods are categorized as mechanical, hydraulic, ultrasonic, microwave, electrostatic, electrodynamic, and electromagnetic fields [22]. Some of these methods are applicable at the lab scale and are still in the research stage. In the following subsections, the conventional methods are discussed.

2.3.2.1 *Mechanical removal*

The mechanical removing of asphaltenes uses mechanical instruments to physically dislodge and remove the solid deposits in the wellbore. Drilling, scraping, and brushing with any combination of bits, blades, and bristles are among the mechanical methods of asphaltene deposition removal which are usually implemented after a halt in the production (or process) [5]. To select an efficient method, capital and operating costs need to be optimized for a well depending on the depth, length, and severity of the damaged region. Plugged pipes due to asphaltene depositions are treated using plug-like devices, known as pigs. The pigs can be equipped with features such as bristles and scraper blades to result in higher efficiency, and also sensors to diagnose the status and severity of blockage by asphaltenes through solid accumulation and leak detections [5]. Although the mechanical methods of asphaltene removal are costly and risky, they are preferred to chemical methods as they minimize the health and environment hazards that are associated with chemicals.

2.3.2.2 Hydraulic

Hydraulic treatment methods of asphaltene deposition removal use fluid pressure to remove deposits. A reinforced wire line is commonly used for cleaning vertical wells, in a downhole operation. However, the invention of coiled tubing in the 1980s made the in-situ operations more convenient and economic due to less assembly time and more capabilities. One of the most important benefits of coil tubing compared to wire line is its capability to treat horizontal wells without stopping production. Thus, coil tubing can be used as hydroblasting, to convey a high-pressure wash nozzle to dislodge the depositions. This technique combines concentric coiled tube and conveyed tool with other features such as jetting and vacuuming capabilities [5].

2.3.2.3 Ultrasonic

Ultrasonication is commonly used for cleaning and removing rusts. Mechanisms of cavitation and bombardment with ultrasonic waves enable this method to be used for cleaning solid deposits. Cavitation releases heat and energy, which increases temperature that eases the dislodging of deposition by loosening them. Moreover, sound speed is different in various media based on their characteristics such as density. Different sound speeds in deposited asphaltene and the surface on which asphaltene adhered will initiate the force that causes asphaltene removal. Both mechanisms have a direct relationship with ultrasonic power and intensity. Moreover, the ultrasonic waves can dissolve the asphaltene deposits. This phenomenon has been verified with the micromodel tests [24].

2.3.2.4 Laser technology

Laser technology has barely been used in the oil industry. Moavenzadeh et al. [25] employed a low power-density laser to weaken and dislodge stones without cracking or breaking them. Hence, the same mechanism is extended to dislodge the asphaltene depositions, and this applicability is verified with the flow assurance problems in porous media [16].

2.3.2.5 Thermal methods

These methods include injecting hot oil, water, steam, and heat-liberating chemicals as well as the use of a downhole heater [16,26,27]. In the hot fluid injection, the main process is to dislodge and dissolve the deposits, but the dissolved particles can be deposited elsewhere. Thus, the operators should make sure the well is totally clean after the process [16,28]. Downhole heaters can be used as a permeant heat source to melt the asphaltene depositions in the wellbore or tubing [16,29]. In the heat-liberating chemical process, a mixture of ammonium chloride and sodium nitrate along with a buffer will be injected into the damaged area; the buffer delays the process [16].

2.4 Common industrial practices

The best asphaltene removal and inhibition treatment techniques are case sensitive that should be chosen based on the treatment costs, the severity of the problem, and the characteristics of the crude oil and reservoir/well. For instance, in the north Aegean Sea, Prinos field, xylene was selected to clean damages by adding a second string to the well and injecting the solvent through a nonproducing string [30]. In the ultra-deep Villafortuna-Trecate, toluene wash was implemented to remove the deposited asphaltene [31]. In Ventura Avenue field, California, solvent wash by toluene was not found efficient, while a mechanical removal technique provided a superior performance [30,32]. An oil well in Venezuela faced a severe asphaltene deposition problem and required cleaning every 7 months. The cleaning involved xylene solvent soak and subsequent mechanical methods such as scraping. The asphaltene treatments impose additional costs to production, on top of the production loss due to frequent well shut-in periods [33]. Jeffries-Harris and Coppel [3] claimed that the use of a surfactant as cosolvent of common solvents such as xylene improves the asphaltene removal efficiency. In asphaltene removal trials in damaged Nebraska wells, scraping was not beneficial while xylene wash showed some effectiveness and the application of solvent and cosolvent, chemical A, had the most efficiency (see Table 2.1) [34].

The Hassi Messaoud field in Algeria suffered from oil deposition in the tubing. First, scraping and solvent washing were used to treat the asphaltene

Table 2.1 Comparison of asphaltene removal methods' peak production, Nebraska wells [34].

Well Number	Treatment	BOPD Before Treatment	BOPD After Treatment
1	Chemical A	32	80
2	Chemical A	8	33
3	Chemical A	38	80
4	Xylene	80	85
5	Xylene	80.5	83
6	Scraper	22	22
7	Scraper	12	8

deposition. However, after several replications, changing the operating conditions was found more efficient to treat asphaltene problem in the tubing [35]. Similarly, in Lake Maracaibo, Venezuela, optimizing the production operating conditions was found more effective than chemical and mechanical treatments [35].

To avoid asphaltene deposition, chemical inhibitors can be applied using downhole capillary injection or injection-soak-production which are examples of continuous injection and cyclic injection, respectively. In the first method, a capillary tube is installed in the well to continuously inject the inhibitors in the producing well. In the second method, the well would be shut down, and the inhibitors will be injected into the damaged area where it is allowed to soak. Then, the well will be reopened to produce the removed asphaltene. In the second case, the inhibitor should adhere to the rock surface and extend the durability of asphaltene inhibitor effect. Based on the location of the asphaltene depositions, both methods can be practical. Moreover, based on the deposition severity and economic factors, the best type and dosage of inhibitor should be designed [4]. In Alberta, Canada, some wells suffer from asphaltene deposition that require frequent solvent wash (every 16 h); in this case, continuous downhole capillary injection was more beneficial and up to 80% cheaper than the cycle of injection (solvent, or inhibitor), soak, and production method. The oil production of well in Southern Louisiana which used to be treated frequently with xylene wash, was extended by more than 5 months when cyclic injection method was implemented. A well in the Permian Basin with asphaltene problem was treated with hot water, hot oil and acid injection, and switched to cyclic injectionmethod. The results showed longer durability and cost-effectivity of the soak method compared to the previous treatment ways [36].

Despite the success of chemical inhibitors in treating asphaltene damage problems, they impose extra cost to the operating costs. In the Middle East, the chemical additive injection costs $31,000–$46,000 per well per year, and for a 10,000 bbl/day production in Gulf of Mexico, the injection of inhibitors costs $330,000–$390,000 per well per year. Therefore, exploring a cost-effective method of asphaltene deposition treatment is of great economic importance [5].

2.5 Theoretical and practical challenges in controlling asphaltene deposition/precipitation

Depending on the crude oil properties and production operating conditions, the asphaltene deposition can be severer. In such cases, physical and chemical treatment methods can be combined to effectively reduce the asphaltene problem. An economic assessment to control the asphaltene deposition problem is important to ensure optimal operation of the oil wells [5]. For instance, repeating the soaking cycles and longer soaking time result in a better performance, although repetition can reduce the effect of acid treatment. The industrial sector requires clean, effective, and economical methods for resolving asphaltene problems with minimum environmental impacts [10]. Although utilizing aromatic solvents is an excellent method to revive asphaltene-damaged wells, these chemicals are flammable and carcinogenic, which are hazardous to the environment, ecosystem, and human health [6]. The solvent wash method needs an accurate wireline implementation to treat the damaged zone; also, cyclic injection relies on the appropriate soaking time and inhibitor effectiveness. Effective chemical inhibitors have superiority over the conventional solvent treatment methods as they can increase the production rate with less frequent treatment demand [33]. However, injecting a new fluid into a hydrocarbon reservoir may carry additional consequences. Thus, the compatibility between the injected fluid and the in-situ reservoir fluid should be also considered when targeting an asphaltene-related problem [37]. Although the chemical methods provide high efficiency, the problems attributed to their environmental and hazardous footprints motivate researchers and engineers for designing thermodynamic inhibitors. Resins, deasphalted oil, and vegetable oil are potential thermodynamic inhibitors used to tackle the asphaltene problem. Hence, thermodynamic inhibitors are recognized as an alternative treatment method due to their high solubility in crude oil, cheaper investment costs, and easier handling [6]. The thermodynamic inhibitors substantially improve the problems associated with handling and environmental concerns related to the chemicals, but they still have side chemical interactions and incompatibility with reservoir fluid; hence, the physical methods have gradually become more attractive [37]. Physical methods do not have health and hazard problems; however, they are expensive, and the mechanical tools might get stuck in the well, imposing fishing or drilling costs. Some of the mechanical methods may accelerate corrosion or may even damage the casing or tubing and in some cases they cause perforation closure [16,38]. Thermal methods need to be designed, implemented, and maintained; often, they are costly and less efficient than other methods [16,26,27]. The biological methods are slow and are also sensitive to temperature, pH, and salinity; hence, they require a good control system [16,23].

2.6 Final remarks

In the context of solutions to the flow assurance problem by asphaltenes, two strategies are possible. The first strategy focuses on inhibition or mitigation of asphaltene aggregation, to lower the chance of asphaltene precipitation and deposition. The second strategy is removing asphaltenes after its deposition. Generally, the inhibition (or mitigation) of asphaltene deposition is more beneficial compared to treatment of asphaltene problem after its deposition. Different chemical, physical, and biological methods are proposed to treat the asphaltene issue. Chemical methods have shown more promising results, but they have some drawbacks such as environmental issues. The physical methods do not carry environmental concern, but they can damage the production facilities and usually have less efficiency compared to chemical methods. The biological methods are too slow to be economical on the field scale. They are also sensitive to reservoir conditions, causing the failure risk of these methods due to high uncertainties in hydrocarbon reservoirs. The effectiveness of each method varies for different crude oils. Sometimes various methods are combined to use the advantages of each method and increase the treatment efficiency. Therefore, the best treatment method should be studied and developed for each case, and economic factor always should be taken into account.

References

[1] C.-L. Chang, H. Fogler, Asphaltene stabilization in alkyl solvents using oil-soluble amphiphiles, in: SPE International Symposium on Oilfield Chemistry, Society of Petroleum Engineers, 1993.

[2] C.-L. Chang, H. Scott Fogler, Peptization and coagulation of asphaltenes in apolar media using oil-soluble polymers, Fuel Sci. Technol. Int. 14 (1–2) (1996) 75–100.

[3] M.J. Jeffries-Harris, C.P. Coppel, Solvent stimulation in low gravity oil reservoirs, J. Petrol. Tech. 21 (02) (1969) 167–175.

[4] A. Albannay, T. Masuzawa, C.A. Dabbouk, Alternative methods in preventing asphaltene deposition, in: Abu Dhabi International Petroleum Exhibition and Conference, Society of Petroleum Engineers, 2010.

[5] F.M. Vargas, M. Tavakkoli, Asphaltene Deposition: Fundamentals, Prediction, Prevention, and Remediation, CRC Press, 2018.
[6] L.C.R. Junior, M.S. Ferreira, A.C. da Silva Ramos, Inhibition of asphaltene precipitation in Brazilian crude oils using new oil soluble amphiphiles, J. Petrol. Sci. Eng. 51 (1–2) (2006) 26–36.
[7] A.T. Khaleel, M.I.L. Abutaqiya, C.J. Sisco, F.M. Vargas, Mitigation of asphaltene deposition by re-injection of dead oil, Fluid Phase Equilib 514 (2020) 112552.
[8] M. Mullakaev, V. Abramov, A. Abramova, Ultrasonic automated oil well complex and technology for enhancing marginal well productivity and heavy oil recovery, J. Petrol. Sci. Eng. 159 (2017) 1–7.
[9] C. Shi, et al., Application and mechanism of ultrasonic static mixer in heavy oil viscosity reduction, Ultrason. Sonochem. 37 (2017) 648–653.
[10] H. Qiang, T. Donghan, C. Laicheng, T. Shicheng, H. Yongle, Ultrasonic Irradiation Reduces Shengli Heavy Oil Viscosity, Pennwell Publ Co Energy Group, Tulsa, OK, 2017.
[11] H. Hamidi, et al., The effect of ultrasonic waves on oil viscosity, Pet. Sci. Technol. 32 (19) (2014) 2387–2395.
[12] M. Salehzadeh, A. Akherati, F. Ameli, B. Dabir, Experimental study of ultrasonic radiation on growth kinetic of asphaltene aggregation and deposition, Can. J. Chem. Eng. 94 (11) (2016) 2202–2209.
[13] A.N. Tutuncu, R. Joha, An experimental study for removal of near-wellbore asphaltene deposits using ultrasonics, in: SEG Annual Meeting, Society of Exploration Geophysicists, Las Vegas, 2008, pp. 9–14.
[14] I. Najafi, et al., Quantifying the role of ultrasonic wave radiation on kinetics of asphaltene aggregation in a toluene–pentane mixture, Pet. Sci. Technol. 29 (9) (2011) 966–974.
[15] B. Ju, Z. Luan, Z. Wu, G. Lu, A study of removal of organic formation damage by experiments and modeling approaches, in: SPE Asia Pacific Oil and Gas Conference and Exhibition, Society of Petroleum Engineers, 2001.
[16] A.Y. Zekri, S.A. Shedid, H. Alkashef, A novel technique for treating asphaltene deposition using laser technology, in: SPE Permian Basin Oil and Gas Recovery Conference, Society of Petroleum Engineers, 2001.
[17] J. Curtis, Environmentally favorable terpene solvents find diverse applications in stimulation, sand control and cementing operations, in: SPE Annual Technical Conference and Exhibition, Society of Petroleum Engineers, 2003.
[18] K.A. Frost, R.D. Daussin, M.S. Van Domelen, New, highly effective asphaltene removal system with favorable HSE characteristics, in: SPE International Symposium and Exhibition on Formation Damage Control, Society of Petroleum Engineers, 2008.
[19] L. Salgaonkar, A. Danait, Environmentally acceptable emulsion system: an effective approach for removal of asphaltene deposits, in: SPE Saudi Arabia Section Technical Symposium and Exhibition, Society of Petroleum Engineers, 2012.
[20] L. Houchin, L. Hudson, The prediction, evaluation, and treatment of formation damage caused by organic deposition, in: SPE Formation Damage Control Symposium, Society of Petroleum Engineers, 1986.
[21] C. Chen, J. Guo, N. An, Y. Pan, Y. Li, Q. Jiang, Study of asphaltene dispersion and removal for high-asphaltene oil wells, Pet. Sci. 9 (4) (2012) 551–557.
[22] G.A. Mansoori, A. Elmi, Remediation of asphaltene and other heavy organic deposits in oil wells and in pipelines, Socar Proc. 4 (2010) 12–23.
[23] R.A. Almehaideb, A.Y. Zekri, Possible use of bacteria/steam to treat asphaltene deposition in carbonate rocks, in: SPE European Formation Damage Conference, Society of Petroleum Engineers, 2001.

[24] R.R. Dehshibi, A. Mohebbi, M. Riazi, M. Niakousari, Experimental investigation on the effect of ultrasonic waves on reducing asphaltene deposition and improving oil recovery under temperature control, Ultrason. Sonochem. 45 (2018) 204–212.
[25] F. Moavenzadeh, F. McGarry, R. Williamson, Use of laser and surface active agents for excavation in hard rocks, in: Fall Meeting of the Society of Petroleum Engineers of AIME, Society of Petroleum Engineers, 1968.
[26] A. Al-Qasim, A.B. Omar, M. Ghamdi, M. Al-Asker, Asphaltene precipitation: a review on remediation techniques and prevention strategies, in: Abu Dhabi International Petroleum Exhibition & Conference, Society of Petroleum Engineers, 2018.
[27] K. Gharbi, K. Benyounes, M. Khodja, Removal and prevention of asphaltene deposition during oil production: a literature review, J. Petrol. Sci. Eng. 158 (2017) 351–360.
[28] J.H. Pacheco-Sanchez, G.A. Mansoori, In situ remediation of heavy organic deposits using aromatic solvents, in: Latin American and Caribbean Petroleum Engineering Conference, Society of Petroleum Engineers, 1997.
[29] M. Bernadiner, Advanced asphaltene and paraffin control technology, in: SPE International Symposium on Oilfield Chemistry, Society of Petroleum Engineers, 1993.
[30] K.J. Leontaritis, G.A. Mansoori, Asphaltene deposition: a survey of field experiences and research approaches, J. Petrol. Sci. Eng. 1 (3) (1988) 229–239.
[31] M. Galoppini, Asphaltene deposition monitoring and removal treatments: an experience in ultra deep wells, in: European Production Operations Conference and Exhibition, Society of Petroleum Engineers, 1994.
[32] R.N. Tuttle, High-pour-point and asphaltic crude oils and condensates, J. Petrol. Tech. 35 (06) (1983) 1192–1196.
[33] L.M. Cenegy, Survey of successful world-wide asphaltene inhibitor treatments in oil production fields, in: SPE Annual Technical Conference and Exhibition, Society of Petroleum Engineers, 2001.
[34] M. Trbovich, G. King, Asphaltene deposit removal: long-lasting treatment with a co-solvent, in: SPE International Symposium on Oilfield Chemistry, Society of Petroleum Engineers, 1991.
[35] C.E. Haskett, M. Tartera, A practical solution to the problem of asphaltene deposits-Hassi Messaoud field, Algeria, J. Petrol. Tech. 17 (04) (1965) 387–391.
[36] H. Becker, D. Thomas, W. Doddridge, D. McDougall, Asphaltene deposition control using chemical control agents, in: Annual Technical Meeting, Petroleum Society of Canada, 1992.
[37] S.-W. Wong, F. van der Bas, J. Groenenboom, P. Zuiderwijk, Near Wellbore stimulation by acoustic waves, in: SPE European Formation Damage Conference, Society of Petroleum Engineers, 2003.
[38] T. Straub, S. Autry, G. King, An investigation into practical removal of downhole paraffin by thermal methods and chemical solvents, in: SPE Production Operations Symposium, Society of Petroleum Engineers, 1989.

CHAPTER THREE

Fundamentals of chemical inhibitors of asphaltenes

3.1 Chemical inhibitors

The chemical additives that target asphaltene problems are classified as either mitigators or inhibitors. The mitigators are additives that stick to the asphaltene particles, restricting the asphaltene aggregation through which the asphaltene deposition problem is mitigated (or delayed). Addition of mitigators at high concentrations can cause asphaltene self-aggregation that makes asphaltene damage even worse. Inhibitors are defined as additives that change the solution composition, leading to a change of the asphaltene onset point to a more stable condition. The main difference between the mitigators and inhibitors is that the mitigators only postpone the asphaltene precipitation while the inhibitors restrict the flocculation and deposition of asphaltene particles. In other words, the mitigators control the asphaltene deposition, while the inhibitors control the asphaltene precipitation phase. In the literature, these two terms are used interchangeably. Here, we name both additives as chemical inhibitors.

3.2 History/background of inhibitors

In 1969, inhibitors were first used as cosolvents during the solvent wash process when Jeffries-Harris and Coppel [1] used aromatic solvents, amines, alcohols, and carbon disulfide (as cosolvents of diesel oil) to remove asphaltene from clay sand packs. In this process, butylamine was recognized as the most effective cosolvent for asphaltene removal [1]. In 1975, different amines were tested as cosolvents of xylene for asphaltene dissolution. Ethylenediamine and butylamine were found to be effective in dissolving

Asphaltene Deposition Control by Chemical Inhibitors
https://doi.org/10.1016/B978-0-323-90510-7.00004-5

asphaltene based on the exposure time and the cosolvent concentration [2]. Newberry and Barker [3] investigated various surfactants such as oleic acid, phenol, benzoic acid, pyridine, dodecylbenzenesulfonic acid (DBSA), and octanoic acid as cosolvents of xylene. A mixture of xylene and DBSA featured the highest solubility; among sulfonic acids with different alkyl tail lengths, tetradecylbenzenesulfonic acid was the most effective cosolvent. Sutton [2], and Newberry and Barker [3] compared the effectiveness of DBSA and amines to dissolve asphaltenes, and concluded that DBSA is more effective compared to amines [3]. The mechanistic studies showed that cosolvents decrease the extent of asphaltene deposition in a different mechanism compared to the solvent's mechanism. The cosolvents surround the asphaltene particles and keep them suspended in the oil phase, which is similar to the asphaltene stabilization in crude oil by resins. The cosolvent mechanisms of asphaltene stabilization opened a new path in the asphaltene chemical treatment studies. Since then, investigators have studied and proposed several chemicals to inhibit asphaltene precipitation and deposition.

3.3 Types and structures of chemical inhibitors

Based on the structure and inhibitory mechanisms, the chemical inhibitors are classified as kinetic or thermodynamic inhibitors; each of these categories is further divided into other subcategories, as shown in Fig. 3.1. Further discussion on these subcategories is given in next subsections.

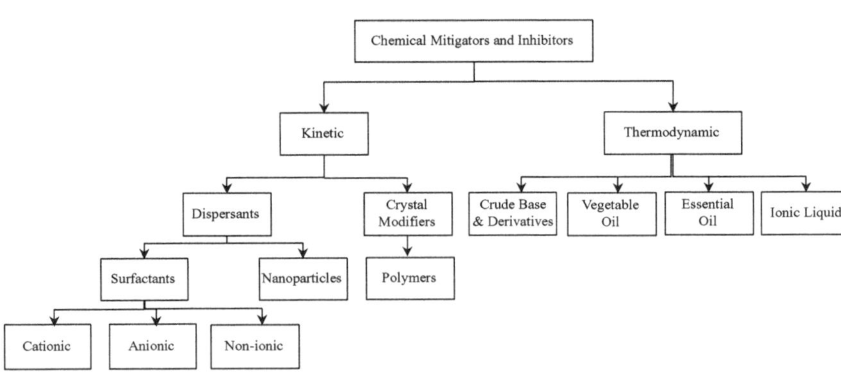

Fig. 3.1 The classification of chemical additives.

3.3.1 Kinetic inhibitors

This section focuses on kinetic inhibitors, which are often used in low dosages. Kinetic inhibitors surround the asphaltene molecules, and through steric effects hinder the aggregation of asphaltene particles. Dispersants and crystal modifiers are two important classes of kinetic inhibitors. Dispersants such as surfactants and nanoparticles will be dispersed around the asphaltene molecules and inhibit asphaltene aggregation. Crystal modifiers, such as polymers, modify the surface chemistry of the asphaltene molecules through their functional groups to limit the growth of asphaltene aggregates. In some research studies, the type and structure of commercial inhibitors are not disclosed; we named them as commercial inhibitors.

3.3.1.1 Surfactant

The surfactants are divided into ionic and nonionic classes; the ionic surfactants are either anionic or cationic. Anionic surfactants have a negative charge due to their anionic functional head in their structure. Sulfonic acid, carboxylic acid, and sulfonate are common functional groups in anionic surfactants. Surfactants were used as cosolvents to improve the dissolution of asphaltene in common solvents more effectively [2, 3]. DBSA has a sulfonic acid moiety attached to an aromatic ring. It has shown superiority compared to other cosolvents as tested in many studies [4–30]. Different inhibitors based on alkylbenzenesulfonic acid are also investigated, and their inhibitory effects are compared to conventional DBSA [13, 31–34]. The DBSA-based salts, which have sodium instead of the hydroxyl group, such as sodium dodecylbenzene sulfonate [9, 35] and sodium octylbenzenesulfonate [9], are also considered as asphaltene inhibitors. Similarly, sodium dodecyl sulfate (SDS) [13, 36], triethanolamine lauryl ether sulfate [31], sodium lauryl ether sulfate [31], and dioctyl sodium sulfosuccinate (AOT) [37, 38] are also potential anionic surfactants for asphaltene inhibition; they have a sulfonate group and no aromatic rings. These inhibitors are used in CO_2 injections and oil production to limit asphaltene deposition. Wiehe and Jermansen [39] concluded that the sulfonic acid-based functional groups inhibit asphaltene deposition better than the carboxylic, hydroxyl, and amine groups. They suggest IsoC30 naphthalenesulfonic acid as the best inhibitor after sensitivity analysis on the functional head group, the number of aromatic rings, number of carbon in alkyl tails, carbon number, and tail branching degree [39].

Surfactants with a carboxylic head group and an alkyl tail are also anionic inhibitors. In 1969, Jeffries-Harris used both acetic acid and isopropanol alcohol as cosolvents for diesel oil to mitigate asphaltene precipitation [1]. Caprylic acid [3, 4, 18] and oleic acid [3, 18] are other anionic surfactants that have been used as inhibitors with xylene. Lauric acid [40], palmitic acid [4], and stearic acid [41] are other alternatives with similar functional head to caprylic acid [3, 4, 18] but different alkyl tails. Linoleic acid [4] has the same functional group as oleic acid [3, 18] while it has two π bonds in its tail.

An anionic surfactant may have more than one functional group. To evaluate the effectiveness of additional functional groups, the performances of benzoic acid [3, 13, 36, 42], octylbenzoic acid [32], salicylic acid [4, 13, 36, 42], phthalic acid [13, 42], and sebacic acid [4] are compared. These inhibitors have a carboxylic acid functional group and a benzene ring (except for sebacic acid). Benzoic acid and octylbenzoic acid do not have additional functional groups, while phthalic acid [13, 42] and sebacic acid [4] have an additional carboxylic acid functional group, and salicylic acid [4, 13, 36, 42] has an additional hydroxyl functional group.

Cationic surfactants have a positive charge due to the functional groups in their heads. Although the cationic inhibitors have not attracted much attention similar to anionic inhibitors, promising inhibitory results are also obtained with the cationic inhibitors and they are under investigation. Since amines were shown to have acceptable efficiency as cosolvents in past studies, ammonium-based materials including cetyltrimethylammonium bromide (CTAB) and dodecyltrimethylammonium bromide (DTAB) are suggested as potential cationic inhibitors. Both CTAB and DTAB have the same functional head but have different lengths of alkyl tail. CTAB was tested for the first time in 1999 as an inhibitor, and its effectiveness was compared with other types of inhibitors [35]. The promising inhibitory effect with CTAB has motivated researchers to conduct further research on cationic asphaltene inhibitors [20, 36].

Moreover, the zwitterionic gemini surfactant has been used as an asphaltene inhibitor. This type of surfactant is composed of two hydrophilic head groups linked by a spacer with two hydrophobic tail groups attached to each hydrophilic group. The functionality of these materials as inhibitors has been recently uncovered. To the best of our knowledge, the only inhibitor of this type is reported by Alcázar-Vara et al. in 2015, who introduced geminal zwitterionic liquid as an inhibitor [43, 44].

Phenols, alcohols, ethoxylated phenols, ethoxylated alcohols, and amines are categorized as nonionic inhibitors, as they carry zero net molecular charge. Researchers have studied phenols with different lengths and alkyl tail structures. Phenol [3, 13] is considered as the simplest nonionic inhibitor because it includes an aromatic ring and a hydroxyl functional group and no alkyl tail. For example, cresol [1, 26, 29, 41], ethylphenol (EP) [9, 26, 29], butylphenol (BP) [9, 29], hexylphenol (HP) [9, 26, 29, 45], octylphenol (OP) [9, 29, 40, 45, 46], nonylphenol (NPH) [5–7, 9, 11, 13, 15, 16, 23, 25, 26, 30, 41, 42, 47–53], and dodecylphenol (DP) [9, 11, 14, 23, 26, 29, 45–47, 54] have similar functional heads to phenol with different lengths of alkyl tails. Phenols such as *sec*butylphenol (SBP) [9, 11, 26, 29], *ter-t*butylphenol (TBP) [11], and *tert*octylphenol (TOP) [11, 26, 29] are compared to study the effect of branches in the alkyl tail. Furthermore, to study the effect of additional functional groups (for phenols), creosol [15, 25, 55] and heptyloxyphenol (HOP) [26, 29] are compared in terms of inhibition performance. The difference between these two components is in their alkyl tail and number of alkyl tails attached to their benzene rings. In the literature, nonylphenol and dodecylphenol are the most used chemicals in the category of nonionic inhibitors.

Alcohols also have a hydroxyl functional group, and unlike phenols, they do not have an aromatic ring in their molecular structure. Methanol [1], 1-octanol [28, 49], and dodecanol [56] are alcohols with different lengths of alkyl tails and are used as nonionic inhibitors. Moreover, alcoholic isomers like 2-ethyl-1-hexanol [28] and isopropanol [1], which have branches in their alkyl tails, show some inhibitory effects. Benzyl alcohol [28] is another type of alcohol that has a benzene ring in its tail. Recently, Triton X-100 was tested as a nonionic inhibitor that features some molecular structure similarity to benzyl alcohol [36].

The other common nonionic surfactants are ethoxylated phenols and ethoxylated alcohols, which exhibit inhibitory effects [49]; for example, ethoxylated nonylphenol with varying degree of ethoxylation is studied in the literature [4, 11, 13, 18, 26, 29, 30, 35, 47, 49, 56]. In the ethoxylation reaction of nonylphenol, the hydrogen on the hydroxyl group is substituted by the ethylene oxide group where the number of ethylene oxide groups is related to the degree of ethoxylation. In addition, ethoxylated nonylphenols with phosphonate group substitutions are used as nonionic inhibitors in asphaltene inhibition [49]. The alcohols and their ethoxylation are compared together to check whether the ethoxylated groups can improve the alcohol's inhibitory effects [31]. The nonionic inhibitors with additional functional groups such as 4-dodecylresorcinol (DR) [7, 14, 26, 29, 57], sorbitan monooleate (SM) [30], and PL [13] have extra hydroxyl groups in their molecular structures.

Amines are the other important category of the nonionic inhibitors with appreciable inhibitory effects [1, 2]. Complex amines such as hexylamine [1, 28], dodecylamine (Noram C) [40, 49], and hexadecylamine [41] are studied; they feature the same functional heads and different lengths of alkyl tail. Diethanoldodecylamine (DDA) [30] and coconut diethanol amide [31, 33] consist of two ethanol groups attached to nitrogen, which contain different lengths of the alkyl tail. Recently, Chavez-Miyauchi et al. proposed a new class of amino alcohols with a different alkyl tail [58]. Table 3.1 summarizes the structure and molecular weight of various surfactant types that have been used as asphaltene inhibitors so far.

3.3.1.2 Nanoparticle inhibitors

Asphaltene particles can be adsorbed onto different surfaces, including minerals (clay and rock), glasses, metals, metal oxides, and carbons [59]. The adsorption of asphaltene can change the surface wettability in porous media, tubing, and surface facilities; however, the adsorption of asphaltene on a suspended surface can reduce the asphaltene self-aggregation. Nanoparticles are regarded as potential asphaltene inhibitors if they stay dispersed and suspended for enough time. Their high surface-to-volume ratio (specific surface) and small size enable them to adsorb more asphaltene particles and percolate through porous media [60]. The interparticle forces between the nanoparticles and asphaltenes' functional groups (such as carboxylic, pyrrolic, pyridinic, thiophenic, and sulfite) lead to high adsorption capacity [61]. Fritschy and Papirer [62] were pioneered regarding the investigation of metal oxide nanoparticles as asphaltene inhibitors. They used silica (SiO_2), alumina (Al_2O_3), magnesium oxide (MgO), zinc oxide (ZnO), titanium dioxide (TiO_2), and kaolin to investigate the adsorption mechanisms in a nanoparticle-asphaltene system [62]. Alboudwarej et al. studied the adsorption of Athabasca and Cold Lake asphaltene on three metal nanoparticles, including stainless steel (304L), iron, and aluminum at different temperatures and compositions [63]. After 2005, nanoparticle application became a hot topic, and many researchers such as Dudasova et al. [64], Hosseinpour et al. [65], Nassar et al. [66], and Franco et al. [67] studied asphaltene adsorption on nanoparticles as an inhibitory index. As a result, the following hydrophilic and hydrophobic nanoparticles are introduced and

Table 3.1 Different types of surfactant which are used as asphaltene inhibitor in the references.

Inhibitor	Chemical structure	Mw (g/mol)	Ref.	Inhibitor	Chemical structure	Mw (g/mol)	Ref.
Anionic surfactants							
Dodecylbenzenesulfonic acid (DBSA)	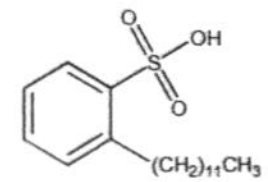	326.5	[4–30]	Branched DBSA	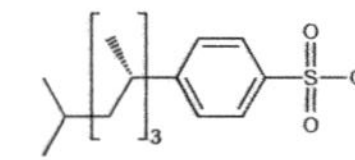	321	[31, 33]
Linear alkylbenzenesulfonic acid	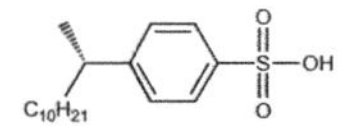	326.5	[13, 31–33]	i-C_{30} Naphthalene sulfonic acid	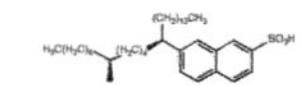		[39]
Dioctyl sodium sulfosuccinate	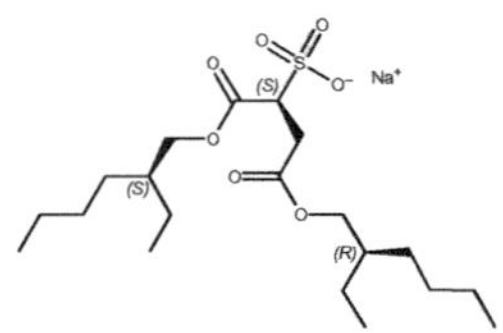	444.6	[37, 38]	Sodium octylbenzenesulfonate- R = [C_8H_{17}] Sodium dodecylbenzenesulfonate- R = [$C_{12}H_{25}$]	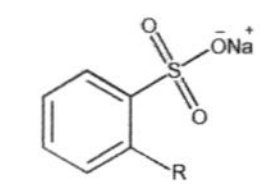	292.4–348.5	[9] [9, 35]
Triethanolamine lauryl ether sulfate, R = [$C_{12}H_{25}$] and R' = [$(CH_2)_2OH$]	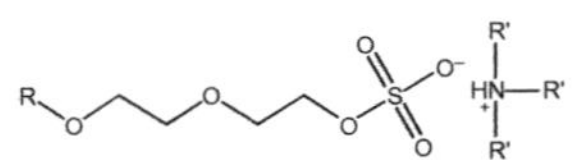	415.6	[31]	Sodium dodecyl sulfate (SDS)	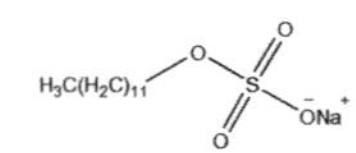	228.4	[13, 36]
Sodium lauryl ether sulfate	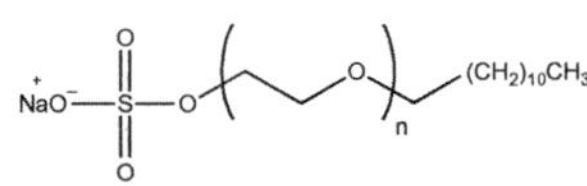	496.7	[31]	Sebacic acid	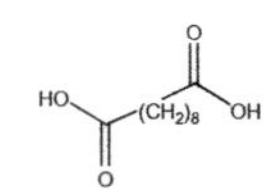	202.3	[4]
Benzoic acid, R = [N/A] Salicylic acid, R = [OH] Phthalic acid, R = [COOH]	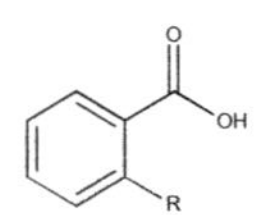	122.1–234.3	[3, 13, 36, 42] [4, 13, 36, 42] [13, 42]	Acetic acid R = [CH_3] Caprylic acid [C_7H_{15}] Lauric acid [$C_{11}H_{23}$] Palmitic acid [$C_{15}H_{31}$] Stearic acid [$C_{17}H_{35}$]	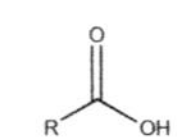	60–284	[1] [3, 4, 18] [40] [4] [41]
Linoleic acid	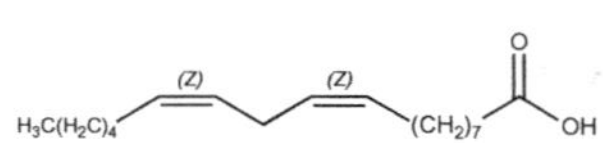	280.4	[4]	Oleic acid (fatty acid)	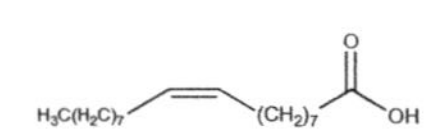	282.5	[3, 18]

Continued

Table 3.1 Different types of surfactant which are used as asphaltene inhibitor in the references—cont'd

Inhibitor	Chemical structure	Mw (g/mol)	Ref.	Inhibitor	Chemical structure	Mw (g/mol)	Ref.
Nonionic surfactants							
Phenol R = [N/A]	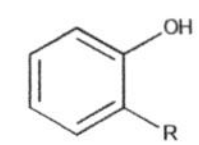	94.1 – 262.4	[3, 13]	Renex 18 (nonylphenol ethoxylated), R = [H], n = 1.8	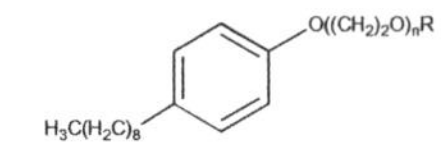	299.2	[4, 18, 49]
Methylphenol [CH_3]			[1, 26, 29, 41]	Renex 20		308	[13, 26, 29]
Ethylphenol [C_2H_5]			[9, 26, 29]	Renex 40		396	[4, 13, 18, 49, 56]
Butylphenol [C_4H_9]			[9, 29]	Renex 50, 70, 110		440–704	[56]
Hexylphenol [C_6H_{13}]			[9, 26, 29, 45]	Renex 60		484	[11, 30, 47]
Octylphenol [C_8H_{17}]			[9, 29, 40, 45, 46]	Renex 100		660	[4, 18, 35, 49]
Nonylphenol (NPH) [C_9H_{19}]			[5–7, 9, 11, 13, 15, 16, 23, 25, 26, 30, 41, 42, 47–53]	Renex 200 Renex 230,300 Renex 1000		1100 1232–1540 4620	[35] [49] [4, 35, 49]
Dodecylphenol [$C_{12}H_{25}$]			[9, 11, 14, 23, 26, 29, 45–47, 54]	D-22, R = [$PO(OH)_2$], n = 1.8		379.2	[49]
sec-Butylphenol	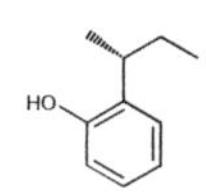	150.2	[9, 11, 26, 29]	tert-butylphenol	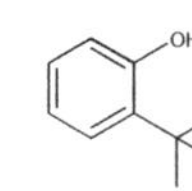	150.2	[11]
tert-Octylphenol	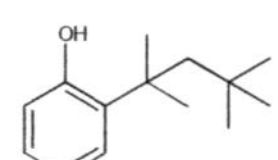	206.3	[11, 26, 29]	heptyloxyphenol	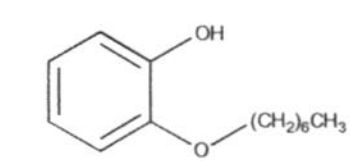	208.3	[26, 29]
2-Methoxy-4-methylphenol (Creosol)	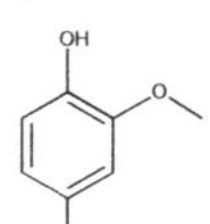	138.2	[15, 25, 55]	Triton X-100	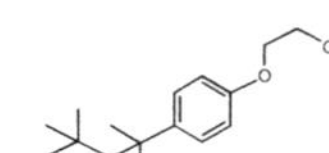	250.4	[36]
1-Octanol R = [C_8H_{17}] Methanol R = [CH_3] Dodecanol R = [$C_{12}H_{25}$]		32–186.3	[28, 49] [1] [56]	Benzyl alcohol	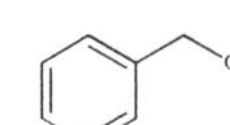	108.1	[28]
2-Ethyl-1-hexanol	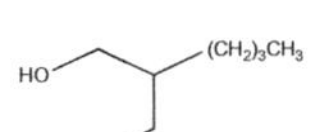	130.2	[28]	Ethoxylated fatty alcohol 9 mol	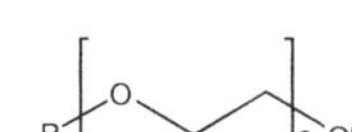		[31]

Table 3.1 Different types of surfactant which are used as asphaltene inhibitor in the references—cont'd

Inhibitor	Chemical structure	Mw (g/mol)	Ref.	Inhibitor	Chemical structure	Mw (g/mol)	Ref.
Isopropanol	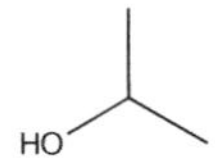		[1]	4-dodecylresorcinol (DR)	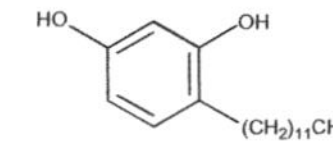	278.4	[7, 14, 26, 29, 57]
Sorbitan monooleate (SM)	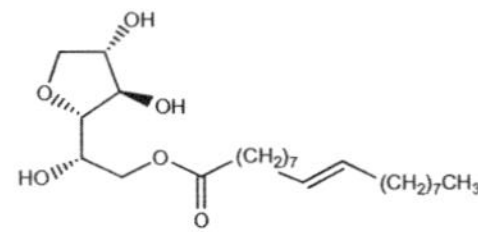	428.6	[30]	Dodecyl-2-pyrrolidinone	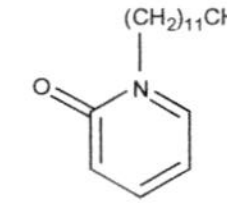	253.4	[30]
Diethanoldodecylamine	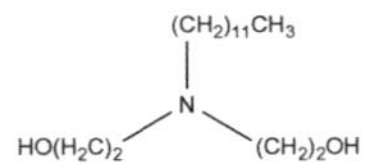		[30]	Morpholine	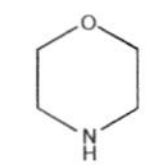	87.1	[2]
Cyclohexylamine	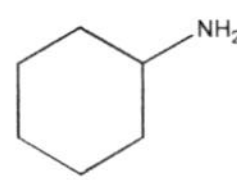	99.2	[2]	Coconut diethanol amide $R = [C_nH_{2n+1}]_{n=7,9,11,15,17}$	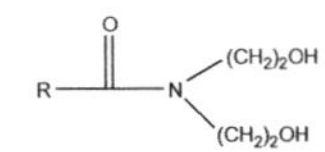		[31, 33]
Butylamine $R = [C_4H_9]$	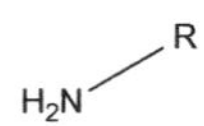	73.1–241	[1, 2]	Monoethanolamine		61.1	[2]
Hexylamine $[C_6H_{13}]$			[1, 28]				
Dodecylamine (Noram C) $[C_{12}H_{25}]$			[40, 49]	Diethylamine	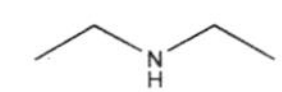	73.1	[2]
Hexadecylamine $[C_{16}H_{33}]$			[41]				
Diethylenetriamine	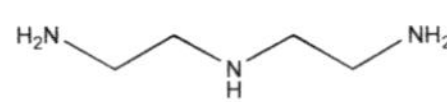	103.2	[2]	Dimethylaminopropylamine	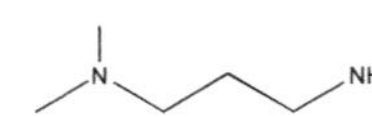	102.2	[2]
N-aryl amino alcohols $R = [C_nH_{2n+1}]_{n=6,10,16}$ $R_2 = H, B(OH)_2$	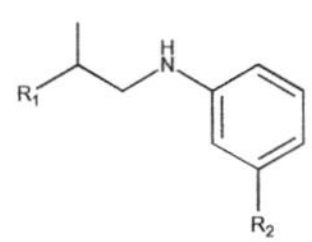		[58]	Diethylaminopropylamine	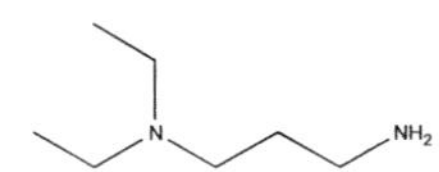	130.2	[2]
Cationic surfactants							
Cetyltrimethylammonium bromide (CTAB)	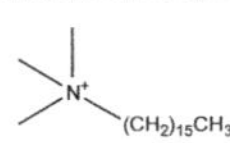	364.5	[35, 36]	Dodecyltrimethylammonium bromide (DTAB)	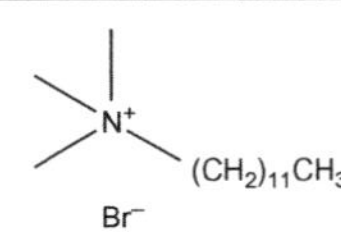	308.4	[20]

tested: kaolin [64], calcium carbonate ($CaCO_3$) [64], barium sulfate ($BaSO_4$) [64], iron sulfide (FeS) [64], silica gel [67, 68], silica (SiO_2) [64, 67, 69, 70], titanium dioxide (TiO_2) [64, 66, 70], iron (II and III) oxide (Fe_3O_4) [60, 64–66, 71, 72], magnesium oxide (MgO) [65, 66, 69], aluminum oxide (Al_2O_3) [60, 61, 67, 69, 73–77], nickel oxide (NiO) [60, 65, 66, 71, 78], tungsten oxide [65], zirconium oxide [65, 70], cobalt oxide [66], and calcium oxide [66]. Furthermore, in these comprehensive studies, nonmetallic nanoparticles were investigated to compare with the popular ones [65, 67].

It is possible to engineer the properties of nanoparticles (such as surface coatings, size distribution, and specific surface) and improve their adsorption capacity. The coating mainly changes the nanoparticle's surface affinity, which can increase the adsorption of asphaltene particles on the surface. The nanoparticle's size and the specific surface can be changed by controlling the temperature, pressure, and time of their synthesis. In 2014, Adams [59] published a review paper on different asphaltene sorbents including metal and metal oxides (with or without further modification); however, this research did not study the nanoparticles from the asphaltene inhibitory perspective [59]. Nassar et al. studied the effect of particle size in Al_2O_3 [77] and NiO [79] micro- and nanoparticles and asphaltene system to study the adsorption capacity at different asphaltene and nanoparticle concentrations. They also investigated the effect of the Al_2O_3 surface acidity as another important parameter [61]. Cortes et al. investigated the adsorption of asphaltene on silica gel, coated with nickel oxide at different levels of temperatures and concentrations [80]. Based on the literature review, Al_2O_3, NiO, SiO_2, and iron (III) oxide are the most effective nanoparticles with application to asphaltene inhibition. Table 3.2 summarizes the nanoparticles and their effective features in asphaltene aggregation inhibition.

3.3.1.3 Polymeric inhibitors

Polymers are macromolecules composed of several chained molecules. Due to the structural similarity of polymers with resins and asphaltenes, they can also be effective in stabilizing asphaltene suspensions. The oil-soluble polymers like poly(maleic anhydride l-octadecene) (POM) [54] are suggested as potential asphaltene inhibitors; their effectiveness is compared with anionic and nonionic inhibitors [23]. Moreover, polymer-based surfactants can be considered as suitable asphaltene inhibitors. Of examples are nonylphenolic resin (NPR) [48] and dodecyl phenolic resin (DPR) [54]. Similar to the discussion on functional groups, the degree of polymerization highly affects the polymer inhibitory effectiveness. Therefore, polymers with the same

Table 3.2 Characterizations of nanoparticles which are considered as asphaltene inhibitors in previous studies.

Inhibitor	<D>[a] (nm)	S[b] (m^2/g)	Physical properties[c]	Ref.	Inhibitor	<D>[a] (nm)	S[b] (m^2/g)	Physical properties[c]	Ref.
Aluminum	10–14	2.37		[63]	Iron	<40	0.43		[63]
Stainless steel (304L)	<150	0.15		[63]	Kaolin	100–4000	19.7	Plate-like or lamellar, pH = 2	[64]
$CaCO_3$	70	19.9–76.5	Rhombic, pH = 5 (nanofluid)	[64, 65]	FeS	2000	6.9	Sp, pH = 4.8	[64]
$BaSO_4$	80–450	6.5	Tabular, Prismatic, pH = 6.3	[64]	SiO_2-i	14	150	Sp, pH = 3.4, hydrophilic	[64]
Al_2O_3	<50	110–205.7	$\rho = 3.65$ (g/ml), Sp	[69, 74]	SiO_2-o	12	150	Sp, pH = 4, hydrophobic	[64]
Alumina	<200–200,000	123.2–232	Mesoporous (52–60 Å average pores), Ac, B, N	[61, 67, 75–77]	γ-Al_2O_3	<50	>39	Near Sp, Ac	[60, 73]
AlNi5/15	35	17.9–69.9		[67]	SiO_2	10–15	130–300	$\rho = 3.65$ (g/ml), Sp, Ac-B	[62, 69, 70, 81]
SNi5/15	90	23.5–36		[67, 80]	Fumed silica gel	7	119.1		[67]
Silica gel	7–240	8–232.3	Crystalline, Am, Ac, N, B	[67, 68, 80]	ZrO_2	9–20	54.3	Sp (uniform), Am, Ac, B	[65, 70]
TiO_2	<25	50–183	Sp (uniform)	[64, 66, 70]	MgO	<50	82–107.8	B, Basicity = 1.15, Sp	[65, 66, 69]
Fe_3O_4	15–30	43–82	Am, Sp, Acidity = 2.04, Basicity = 0.21	[60, 64, 66, 71, 72, 81]	NiO	<100–3000	33–107	Near Sp, Ac	[60, 65, 66, 71, 78, 79, 81]
Fe_2O_3	44	26.4	Am, Acidity = 1.25, Basicity = 0.13	[65]	WO_3		6.1	Ac, Acidity =1.01	[65]
Co_3O_4	<50	41	Am	[66]	CaO	<160	31	B	[66]
Zeolite	16	70.1		[67]	PdNi/Al	35	221.3		[67]
Washed rock	Mesh 60/40	15		[67]	Unwashed rock	Mesh 60/40	15		[67]

[a]D: Average particle diameter.
[b]S: Specific surface area of particles.
[c]Ac: Acid; Am: Amorphous; B: Basic; N: Neutral; Sp: Spherical.

monomer and different degrees of polymerization (or, molecular weight distribution) are tested in the previous studies [35, 82]. Lima et al. used sulfonated polystyrene with different molecular weights for asphaltene inhibition [83]. In the present decade, researchers have become more interested in polymeric inhibitors. Phenol and amine-based polymers are among the most common inhibitors, including poly olefin amide alkeneamine (with average molecular weight of 2000 g/mol) [38], alkylated phenol (4000 g/mol) [37, 38], nonylphenol formaldehyde resins and poly olefin esters [84], poly olefin alkeneamine (2000 g/mol) [12, 37, 85, 86], poly(dodecylphenol formaldehyde) (1900–8100 g/mol) [87], and propoxylated poly(dodecylphenol formaldehyde) (3000–9200 g/mol) [87] based on research studies between 2010 and 2015.

Recently, researchers have designed more effective polymers with a desired functional group that decreases environmental impacts. Chavez-Miyauchi et al. used four different types of aromatic polyisobutenyl succinimides to study the effect of hydroxyl groups, boronic groups, and lengths of alkane chain on asphaltene inhibition performance [88]. After that, environmentally friendly poly(vinyltoluene-*co*-alpha-methylstyrene) was developed, which has functional groups and polarity that favor inhibition [89]. Lately, polyisobutylene succinimide, polyisobutylene succinic ester, and nonylphenol-formaldehyde resin (modified by polyamines) were studied as asphaltene inhibitors [90]. Table 3.3 summarizes the structure and molecular weight of different polymer types that are used as asphaltene inhibitors.

3.3.1.4 Commercial inhibitors

Commercial inhibitors are usually a mixture of different types of inhibitors where the name and composition of the mixture are not disclosed. In some cases, the active components and/or types are disclosed. In the following, we review commercial inhibitors that are used for asphaltene inhibition. Bouts et al. [91] utilized three commercial inhibitors. The first inhibitor was a synthetic anionic surfactant that contained sulfonic, aromatic, and high molecular weight alkyl groups. The second inhibitor was a mixture of mineral oil and polymeric inhibitor. The third was a mixture of polymeric inhibitor and solvent [91]. Auflem et al. used a mixture of fatty amine and fatty acid [28]. Smith et al. proposed a polymeric-based inhibitor with a protic head and an amine base inhibitor with a polar head [92]. Pereira et al. used aqueous mixtures of ethoxylated nonylphenol and formaldehyde [56]. Barcenas et al. [50, 53] used DAIM-2000 and DAIM-3000 (oxazoline derivatives of polyalkyl or polyalkenyl *n*-hydroxyalkyl) with different concentrations in various

Inhibitor	Chemical structure	Mw (g/mol)	Ref.
Dodecyl phenolic resin	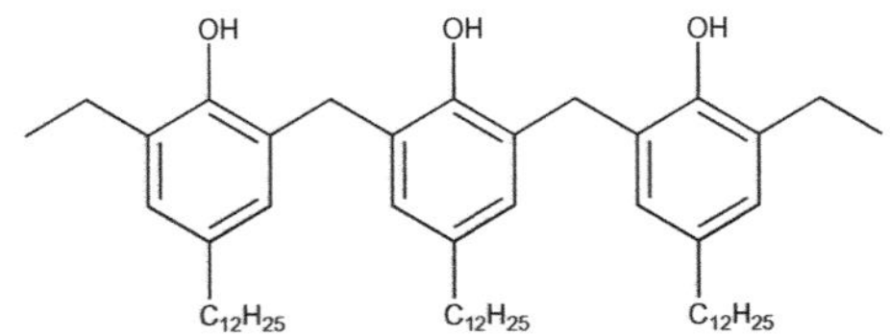	200–2000	[54]
Nonylphenol formaldehyde resins Nonyl phenolic resin	$R = [C_9H_{19}]$, $n = 4,7,20$ $R = [C_9H_{19}]$, $n = 7$ 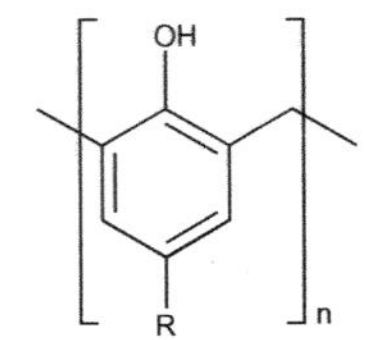	900–4800	[82] [48]
Aromatic polyisobutylene succinimides	$R_1 =$  $R_2 = [H]$, $n = 0,1$; $R_2 = [OH]$, $n = 0$; $R_2 = [B(OH)_2]$, $n = 0$ 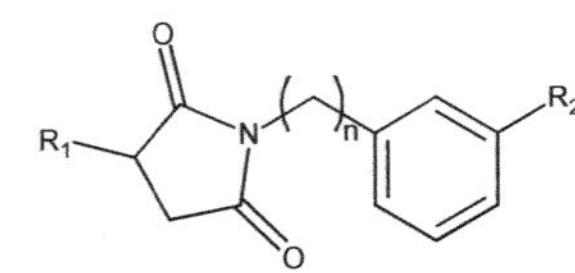		[88]
L31 (n = 17); L64 (n = 30); F108 (n = 48)	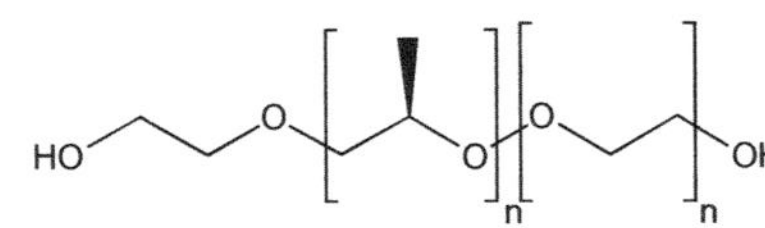	1100–14,000	[35]
Poly(maleic anhydride 1-octadecene) (POM)	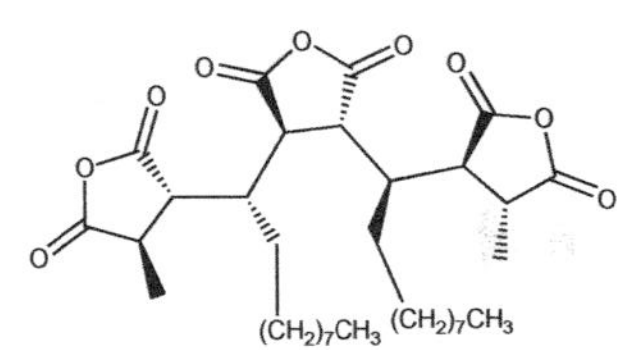	10,000–50,000	[23, 54]

solvents. Chen et al. [93] developed SDJ to disperse and dissolve asphaltene. The SDJ inhibitor is a combination of 10% sodium dodecylbenzenesulfonate, 37% toluene, and 53% isoamyl alcohol [93]. Marcano et al. [24] introduced an inhibitor that consisted of 1%–5% organic sulfonic acid and 1%–5% fatty amine condensate; also an inhibitor was proposed that was a mixture of polymer and hydrocarbon solvents [24]. Joonaki et al. [44] used NCAI as an amphoteric surfactant, which has a benzene ring and a hydroxyl group in its functional head [44]. Kashefi et al. designed SPI as a polymeric inhibitor, which is derived from poly(alkyl phenol formaldehyde) and polyamine (17,700 g/mol) [94]. Wang et al. [95] introduced a biodegradable nonionic polymer with molecular weights ranging from 250 to 10,000 g/mol [95]. Karambeigi et al. introduced IR95, which consists of highly polar and aromatic components [42]. Khormali [96] developed a mixture of polyacrylamide, amphoteric fatty acid, oil-soluble polymers (ethylene copolymer and maleic anhydride), organic sulfates, and 1-hydroxyethane-1, 1-diphosphonic acid (HEDP) as an asphaltene inhibitor [97]. Recently, Subramanian et al. studied fatty alkylamine-based inhibitors to study their interactions with asphaltene molecules [98, 99]. Schantz and Stephenson [100], Aquino-Olivos et al. [101], Marugan et al. [102], Albanny et al. [103], and Yonebayashi and Zhang [104, 105] have used commercial inhibitors with application to asphaltene inhibition without revealing any information about the inhibitor composition or type.

3.3.2 Thermodynamic inhibitors

In this section, we review thermodynamic inhibitors that are used for asphaltene inhibition. These thermodynamic inhibitors are either asphaltene solvents or materials that can be easily blended with the crude oil to change its composition. The thermodynamic inhibitors can be organic, aromatic, and cycloalkane solvents, ionic liquids, vegetable oil, and other crude oil derivates (such as deasphalted oil or resin).

3.3.2.1 Crude base and crude derivative inhibitors

Crude base inhibitors are either crude oil-derivatives or synthetic, which may or may not dissolve asphaltene, but they can be readily blended with the oil to change its composition. Large oil-soluble components include aromatic solvents, deasphalted oils, resins, synthetic, and vegetable oils (and their derivatives). Crude base inhibitors decrease the asphaltene fraction compared to resin and aromatics and shift the asphaltene envelope to inhibit the precipitation and deposition problems. The effectiveness of crude base inhibitors is compared with the kinetic inhibitors in the literature by evaluating their efficiency in reducing asphaltene precipitation and deposition problems. In 1994, Lian et al. [41] used a resin sample with the molecular weight of 800–1300 g/mol. The resin sample included aromatic, hydroxyl, amino, imino, and mercapto functional groups [41]. Alsahhaf et al. used resin and deasphalted oil extracted from Kuwaiti crude oil as inhibitors [7]. Spiecker et al. [106] used resin extracted from California offshore crude, Arab heavy (Safaniya), and Canadon Seco (Argentina) [106]. Goual and Firoozabadi extracted resin from three crude oils with different dipole moments and used them in asphaltene inhibition [8]. Carnahan et al. [107] tested the resins from Boscan, Hamaca, Furrial, and Cerro Negro crude oils. They extracted two types of resins: Resin (I) and Resin (II), where the second one features less polarity [107]. Yakubov et al. used resin and deasphalted oil as asphaltene inhibitors as well [51]. Aside from deasphalted oils and resins, crude and synthetic oil products can also be considered as asphaltene inhibitors. Auflem et al. [28] investigated derivatives of naphthenic acid, including 5-β-(*H*)-cholanoic acid (CHOL), crude naphthenic acid (CNA), 1-napthalenepenanoic acid, decahydro (2C4), and decahydro-2-butyl (C42) [28]. Ghloum et al. tested six different organic inhibitors, which were identified by SARA analysis as inhibitors for four types of Kuwaiti crude oils [108]. Naseri et al. used two crude-oil derived inhibitors, including extracted furfural from oil manufacturing firms (BAR) and heavy aromatics from petrochemical units (HA) [13]. Table 3.4 summarizes few organic inhibitors and their effective features in terms of the inhibition of asphaltene aggregation, which are used research and engineering activities. Aromatics and cycloalkanes are considered as thermodynamic inhibitors. Jeffries-Harris [1] used dicyclopentadiene, as a diesel oil cosolvent with application to asphaltene inhibition. Cyclohexane [1] and decalin [15, 25, 55] with one and two cyclohexane units, respectively, are common cycloalkanes used as asphaltene inhibitors.

From a molecular perspective, the aromatics have resonance bonds in their structure, which allow them to bind to the asphaltene molecules through $\pi - \pi$ intermolecular forces. The asphaltene has multiple aromatic rings in its structure, and it can become stabilized by interacting with a chemical that contains aromatic rings too. Therefore, many chemicals with different number of aromatic rings are studied as inhibitors in research investigations. For instance, benzene [3, 15, 36, 93], toluene [1, 3, 5–7, 13, 15, 17, 25, 27, 31, 33, 50, 51, 55, 109], nonylbenzene [9, 26, 29], and alkyl benzene [49] have a single aromatic ring and a single functional group, and xylene

Table 3.4 Crude base inhibitor and crude derived solvents as thermodynamic asphaltene inhibitors.

Inhibitor	Mw (g/mol)	Physical properties[a]	Ref.	Inhibitor	Mw (g/mol)	Physical properties[a]	Ref.
Crude base							
Dasphalted Kuwait Marrat crude+C_5-C_{15}	384–480	API = 24.7–30.7; R = 0.8–1.2 (wt%)	[7]	Crude oil resins (Boscan, Hamaca, Cerro Negro, Furrial)		API = 8.5–22.7; R/Asp = 1.1–6.6 (wt/wt)	[107]
Diesel, light and heavy cycle oil, also three CI	215–289	H/C = 0.8–1.5 (CI); R = 0.8–1.2 (wt%, CI); Acidic and Basic	[108]	Crude resins (Hondo, Canadon Seco and Arab Heavy)		H/C = 1.3–1.5; R/Asp = 0.25–10 (wt/wt)	[106]
Petroleum resins		API = 33–35; R = 6.5–8.5 (wt%); Asp = 1.3–2.5 (wt%)	[8]	Naphtenic acid (Crude, Fluka, North Sea)	215–400		[28]
5-β-(H)-cholanoic acid	361		[28]	Naphtaleonic acids	238		[28]
Crude resins	800–1300	Include aromatics with hydroxyl, amino, imino, and mercapto groups	[41]	Furfural and heavy aromatics		High aromaticity content	[13]
Crude derivatives							
Toluene	92.14		[3, 5, 7–9, 15, 17, 19, 28, 29, 32]	Benzene	78.11		[3, 15, 36, 93]
				m-Xylene	106.2	$\delta = 18\ (MPa)^{0.5}$	[1, 3, 13, 15, 25, 100, 107]
Cyclohexane	84.16		[1]	1, 2, 4-trimethylbenzene	120.2	$\delta = 17.6\ (MPa)^{0.5}$	[15]
Dicyclopentadiene	132.2		[1]	Decalin	138.3		[15, 25, 55]
Naphthalene	128.2	$\delta = 20.3\ (MPa)^{0.5}$	[15, 25, 36, 41, 55]	Tetralin	132.2	$\delta = 19.4\ (MPa)^{0.5}$	[15, 25, 42, 55]
Quinoline	129.2	$\delta = 22.1\ (MPa)^{0.5}$	[15, 25, 55]	Pyridine	79.1	$\delta = 21.9\ (MPa)^{0.5}$	[3, 15]
Indole (Solid)	117.2	$\delta = 23.2\ (MPa)^{0.5}$	[15, 25, 55]	Benzothiophene	134.2	$\delta = 21.7\ (MPa)^{0.5}$	[15, 25, 55]
Phenanthrene	178.2	$\delta = 20\ (MPa)^{0.5}$	[3, 15]	Nonylbenzene	204.35		[9, 26, 29]

[a]R = Resin wt(%); CI = commercial inhibitor; H/C = ratio of hydrogen-to-carbon; δ: Hildebrand solubility parameter $(MPa)^{0.5}$.

[1, 3, 13, 15, 25, 100, 107] and 1, 2, 4-trimethylbenzene (1,2,4-TMB) [15] have multiple methyl groups attached to the aromatic rings. Naphthalene [15, 25, 36, 41, 55] has two adjacent benzene rings, while phenanthrene [15, 25, 42, 55] has three. These two materials were compared with toluene to capture the impact of existing extra benzene rings [25, 55]. Tetralin [15, 25, 41, 55] was considered as an asphaltene inhibitor and has both benzene rings and cyclohexane. Peramanu, Clarke, and Pruden used aromatic materials with N, S, and O heteroatoms, and hydrogen donor solvents, including pyridine [3, 15], quinoline [15, 25, 55], benzothiophene [15, 25, 55], and indole [15, 25, 55]. Indole and quinoline were used to investigate the effect of basic/nonbasic nitrogen [55]. Table 3.4 presents the popular solvents and their effective characteristics on the inhibition of asphaltene aggregation for the ones that have been used as asphaltene inhibitors in the literature.

3.3.2.2 Vegetable oil and essential oil

Another type of organic inhibitor is vegetable oil. Moreira et al. [110] used cashew-nut shell extract composed of phenolic compounds. They also examined the effect of cardanol (301 g/mol) and polycardanol (900–2300 g/mol), both derived from cashew-nut shell [110]. In 2006, Junior et al. [4] conducted a comprehensive study on a wide type of vegetable oil samples that were derived from nuts, vegetables, and fruits [4]. Safaie et al. [57] compared the effectiveness of hazelnut oil with a surfactant; Monjezi et al. [111] used henna extraction as an asphaltene inhibitor during the pressure decline in a core flood system.

3.3.2.3 Ionic liquids

Ionic liquids were first used as asphaltene inhibitors in 2005 [9]. Ionic liquids have a cationic and an anionic part. Hu and Guo [9] investigated the ionic liquids as inhibitors based on the pyridinium and quinolinium cations with different anions. Pyridinium has a benzene ring while one of its carbon atoms is replaced by nitrogen. Quinolinium is a pyridinium molecule that is attached to a benzene ring. Hu and Guo [9] studied different alkyl groups such as butyl, octyl, and dodecyl as the tail of the cationic section. They studied different anions, including chloride, tetrafluoroborate, and hexafluorophosphate [9]. Boukherissa et al. [109] investigated 1-propylboronic acid-3-alkylimidazolium bromides and 1-propenyl-3-alkylimidazolium bromide with different alkyl tails and 1-butyl-3-methylimidazolium with bromide and chloride as an anion. Imidazolium has a benzene ring with two substituting nitrogens with carbon and, consequently, two alkyl tails.

Didecyldimethylammonium nitrate was used as a nitrogen-based ionic liquid without a benzene ring. Boukherissa et al. studied these ionic liquid inhibitors to analyze the sensitivity of the alkyl tail length, functional head group, and the anion [109]. Since 2014, more attention is paid to these types of inhibitors. The 1-butyl-3-methylimidazolium with chloride and nitrate, and 1-methyl-1*H*-imidazol-3-ium-2-carboxybenzoate were used at different thermodynamic conditions [112]. In 2015, a comprehensive study was performed on ionic liquids, and 1-butyl-3-methylimidazolium with chloride, thiocyanate, and tetrafluoroborate were also investigated to define the effect of the anion. Due to the effectiveness of imidazolium ionic liquids, 1-alkyl-3-methyl imidazolium chlorides with different alkyl groups including ethyl, butyl, hexyl, octyl, decyl, and dodecyl are investigated as well. Dodecyl thiazolium chloride and dodecyl pyridinium chloride were tested to assess the effectiveness of the cation [113]. Recently, more complex ionic liquids are used as inhibitors. In 2016, Atta et al. [114] investigated the effect of synthesized deep eutectic solvent (SDES) and compared its effectiveness with ionic and nonionic surfactants. In 2017, 1-allyl-3-methylimidazolium with cardanoxy, oleate, and abeitate were studied to analyze the effect of alkyl groups, phenoxy groups, and hydrophenancerene groups within the same imidazolium cation [114]. Table 3.5 summarizes the structure and molecular weight of different types of ionic liquids that are used as asphaltene inhibitors in the literature. It worth mentioning that ionic liquids can be considered as a kinetic inhibitor if they are added in low dosage and do not dissolve in crude oil.

3.4 Physical and chemical properties

The crucial characteristics of surfactants are their chemical structure and polarity. The molecular weight of surfactants is reported in studies as a distinct physical property, which is related to their structure. At some point, the lowest limit of the molecular weight of surfactants is proposed to be related to their effectiveness [41], but is not decisively verified. Other physical properties of surfactants such as boiling point and density are sometimes reported, but they are not considered as influential parameters [36]. The structure of polymeric inhibitors is a determinative parameter on polymer inhibitory effect, which is related to the polymerization degree and molecular weight [95]. In nanoparticle inhibitors, surface characteristics, specific surface, and particle size are important parameters. The surface

Table 3.5 List of ionic liquids (ILs) used as thermodynamic asphaltene inhibitors.

Inhibitor	Chemical structure	Mw (g/mol)	Ref.	Inhibitor	Chemical structure	Mw (g/mol)	Ref.
1-Allyl-3-methylimidazolium cardanoxy (AMCO)	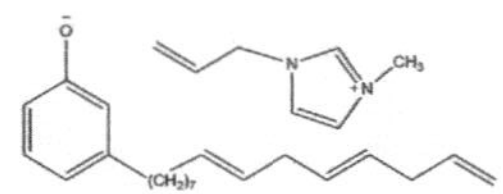		[114]	1-Allyl-3- methylimidazolium oleate (AMO)	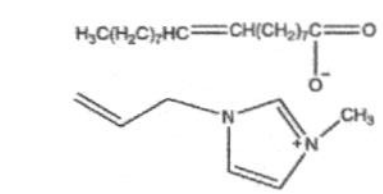		[114]
1-Allyl-3-methylimidazolium abeitate (AMA)	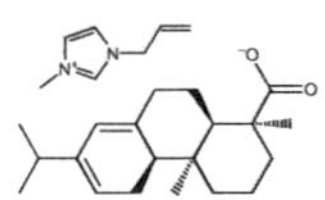		[114]	Butyliso quinolinium chloride	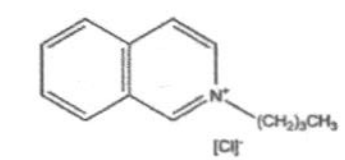	220.7	[9]
(Butyl, octyl, dodecyl) pyridinium chloride	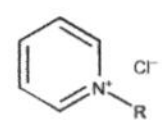 $R = [C_nH_{2n+1}]$ n = 4,8,12	171–284	[9]	Synthesized deep eutectic solvent	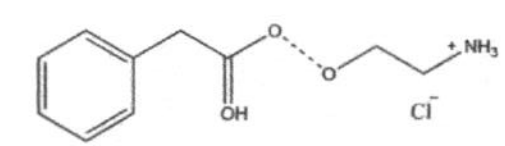		[40]
Dodecyl pyridinium (chloride, tetrafluoroborate, hexafluorophosphate)	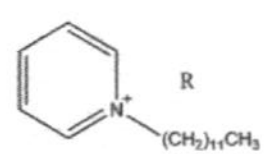 $R = [BsF_4]^-$, $[PF_4]^-$, $[Cl]^-$	342–404	[9, 113]	Didecyl-dimethylammonium nitrate	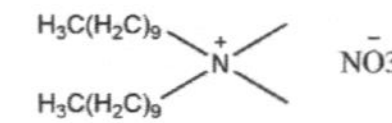	388.6	[109]
1-Propenyl-3-alkylimidazolium bromides	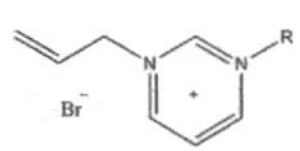 $R = [C_nH_{2n+1}]$ n = 1,2,5,8, 10,12,16		[109]	1-Propylboronic acid-3-alkylimidazolium bromides	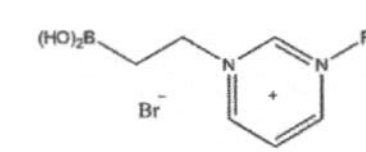 $R = [C_nH_{2n+1}]$ n = 1, 5,8,10,12,16		[109]
1-Butyl-3-methylimidazolium (bromide, chloride, thiocyanate, tetrafluoroborate and nitrate)	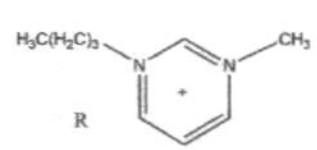 $R = [Br]^-$, $[Cl]^-$, $[SCN]^-$, $[BF_4]^-$, $[NO_3]^-$	174.7–226	[109, 112, 113]	1-(Ethyl, butyl, hexyl, octyl, decyl, dodecyl)-3-methyl imidazolium chlorides	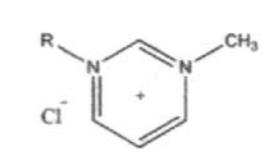 $R = [C_nH_{2n+1}]$ n = 2, 4,6,8,10,12		[113]
dodecylthiazolium chloride	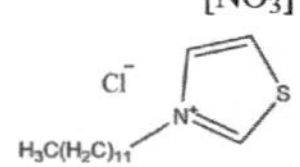		[113]	1-Methyl-1H-imidazol-3-ium-2-carboxybenzoate	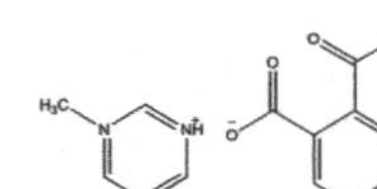		[112]

characteristics govern the adsorption of asphaltene particles with a specific charge and chemistry [61]. The specific surface determines the amount of asphaltene and the small size of nanoparticles enables them to move inside porous media [77]. The shape of nanoparticles is reported in some studies without pointing its relationship with inhibitory effect [60]. The molecular weight, density, and API are the key parameters of organic inhibitors, which affect the compatibility between organic inhibitors and problematic oil [106, 107]. Besides the chemical compatibility, organic inhibitors should have a suitable solubility, resin-to-asphaltene ratio, and hydrogen-to-carbon ratio as determinative parameters [41, 107, 108]. Ionic liquids consist of an anionic and a cationic part, and charge density is an important parameter in their inhibitory effect. Charge density is the molecule's net charge over the ions' size; therefore, charge and size of both cation and anion parts of ionic liquids are essential for their inhibitory effect. In addition, they should have alkyl tails with an optimum length, which is not well defined yet [9, 109].

3.5 Properties and prevention mechanisms of chemical inhibitors

Ideally, the asphaltene inhibitors should be able to adhere to the asphaltene particles and keep them suspended in the solution. The bonding between two chemical materials depends on their structure and chemistry. The structure of the asphaltenes consists of multiple aromatic rings, heteroatoms like sulfur, oxygen, and nitrogen, and metals such as V and Ni. Due to the existence of multiple aromatic rings in the asphaltene structure, they can be attached to each other through $\pi - \pi$ bonding. However, the asphaltene aromatic rings can bind to the chemical inhibitors that capable of bonding with asphaltene polyaromatic core instead of the asphaltene-asphaltene self-aggregation and keep the asphaltene molecules separated. The heteroatoms in the asphaltene structure form hydrogen bonding with chemicals, which have N—H, O—H, and F—H bonds. The heteroatoms and metal particles that induce polarity on the asphaltene molecules allow acid-base interactions with the polar head of inhibitors. These characteristics and bonding properties are related to the asphaltene structure; whereas, the chemical inhibitors should have suitable structure and chemistry such as polar heads and alkyl tails to attach asphaltenes and keep them suspended in dispersive media. Moreover, in the presence of a pipe or rock surface inhibitors adhere to the surface and restrict the asphaltene deposition. The adsorption of asphaltene on the surface depends on inhibitors' charge and surface characterization (and/or characteristics) [34, 42, 43, 97]. In the following sections, further details of the inhibitor chemistry and functional groups based on their types are discussed.

3.5.1 Surfactants

Surfactants consist of two main components: polar heads and alkyl tails. The polar head can consist of one or several functional groups such as the sulfonic, carboxylic, and hydroxylic groups and benzene rings, which can affect the efficiency of the polar head. Wiehe and Jermansen [39] concluded that the optimized number of benzene rings for effective inhibitors is two. The anionic inhibitors have a sulfonic and carboxylic group and usually cause acid-base bonding with asphaltenes, while the nonionic inhibitors usually create a hydrogen bond or van der Waals interaction with the asphaltene due to their O—H and N—H bonds. Negatively charged asphaltenes are dispersed by positive inhibitors, and vice versa [20]. Hence, the cationic surfactants can mitigate the deposition of asphaltene with negative polarity through acid-base bonding. Subramanian et al. [98, 99] used a basic inhibitor and concluded that the acid-base is not the only effective inhibitory mechanism of the ionic inhibitors; on the contrary, Hashmi et al. [21] used an acidic inhibitor and found the acid-base interactions to be the main mechanism for surfactant inhibitors.

Generally, the inhibitory strength of the ionic surfactant inhibitors is usually related to their polar head strength and acidity [4, 49]. The inhibitor concentration is also an important variable in the inhibition effectiveness. The ionic inhibitors can interact with oil compositions other than asphaltene. In this case, the percentage of the inhibitor should be designed in a way that there would be enough free inhibitors to interact with asphaltene, which is related to the crude oil's acid and base number [10]. As an example, Smith et al. [92] used a basic inhibitor and claimed that they are deactivated with oil components that had carboxylic acid. Inhibitors with multiple functional groups can attach to more asphaltene molecules or can create multiple bonds with an asphaltene molecule. For instance, the addition of an extra ether or hydroxyl group on the inhibitor improves its efficiency to inhibit an asphaltene precipitation; the addition of hydroxyl group is more beneficial than the ether group [7, 29].

Other than the attachment of inhibitors to asphaltene molecules, inhibitors should be able to suspend the asphaltenes in the crude oil. Asphaltenes' suspension is related to the hydrophobic part of the chemical inhibitors, known as the alkyl tail. Inhibitors can have one or more alkyl tails carrying the different numbers of alkyl branches in tails. There is a dispute about the number of carbon atoms in the alkyl tail and the degree to create effective steric bonds around the asphaltene. For the single-chain tail, although the asphaltene attraction to the inhibitors with shorter alkyl length is higher [26, 29], different studies proposed an optimum number of carbons in an inhibitor's alkyl tail that is between 6 and 8 carbons [26, 29, 39]. However, in the case of multiple tails, the optimum number increases up to 24 carbon atoms. The reason for optimum number of carbon atoms in the alkyl tail is that if the tail is too short, it cannot keep the asphaltene suspended, and if it is too long, the inhibitor atoms/molecules will associate with each other and form a long (heavy) composition that can deposit in the reservoir. The presence of a branch in the tail can affect the inhibitory effect. There are two hypotheses about the role of branches in the surfactant tails on asphaltene inhibitory performance. In an inhibitor with a short alkyl tail, the branches on tail interfere as a barrier for inhibitors to reach the asphaltene [9, 29]. In an inhibitor with long alkyl tail, the tail branches help the inhibitor be soluble in the hydrocarbon fluid by reducing their self-association and increasing the inhibitors' durability [39]. Therefore, the ineffectiveness of long-chain aliphatic alcohols and alkylbenzenes may be related to their long chain [49].

3.5.2 Nanoparticles

Nanoparticles adsorb the asphaltene molecules and keep them suspended. The efficiency of a nanoparticle for asphaltene inhibition is determined by variables such as size, specific surface (surface to volume ratio), surface acidity, solution pH, and asphaltene characteristics. The nanoparticle size is an important parameter that controls the inhibitory effect, which becomes more critical when the nanoparticle is injected in porous media. By increasing the specific surface area, the chance of asphaltene-nanoparticle adsorption increases. Although the specific surface is directly related to the inhibitory effect [77], a nanoparticle with a high specific surface can have a low inhibitory effect if it does not have effective surface chemistry [69]. Surface acidity of nanoparticles and solution pH along with asphaltene characteristics affect the inhibitory capability of nanoparticles. For asphaltenes with different acidities, nanoparticles or nanofluids with different acidities or pH are effective. For instance, a nanofluid of acidic titanium oxide reduces a asphaltene deposition, while basic nanofluid of the same nanoparticle increases the asphaltene deposition as the result of reduced hydrogen bonding strength in the presence of basic nanofluid [70]. The nanoparticle's surface affinity can be modified with proper chemicals such as acids [60] or nanoparticles [67] to achieve better inhibitory performance [67, 80, 115]. Asphaltene suspension by the additives has prospects: additive attachment to the asphaltene and hindrance of the asphaltene-additives agglomeration. For the case of the asphaltene-nanoparticle suspension, Hosseinpour et al. measured the Gibbs energy of asphaltene adsorption on nanoparticles and found that the asphaltene-nanoparticle has no interest in interacting with the free asphaltenes or nanoparticles [65]. This confirms the ability of nanoparticles to inhibit asphaltene aggregation.

3.5.3 Polymers

Polymers are derived by monomer crystallization. Most of the polymeric inhibitors are based on surfactant inhibitors, and they are expected to have asphaltene inhibition efficiency. Some studies reported the low efficiency of the polymers at low concentrations [23, 54, 110] and also more chance of asphaltene deposition at high concentrations [89]. Therefore, polymers should be added at an optimized concentration. One of the reasons is that the polymers can have multiple functional groups, similar to surfactants, which can cause problems by attaching the two asphaltenes, leading to hetero-coagulation. In this case, the optimal concentration of the polymer inhibitor can sustain the hetero-coagulation—otherwise, they worsen the asphaltene deposition [23, 54, 110]. The molecular weight of the polymers is related to the degree of polymerization, the length of the polymeric tail, and the polarity of their head. Polymers with low molecular weight may not strongly adhere to the asphaltene molecules (to hinder their deposition) due to their low polarity and short alkyl tails, unlike high molecular weight polymers [95].

3.5.4 Organic inhibitors

The organic inhibitors are mainly composed of resin components that inhibit asphaltene deposition by disrupting the $\pi - \pi$ interaction between the asphaltene molecules [106]. The organic inhibitor efficiency is related to their resin fraction and polarity [108]. Although resins and deasphalted oils are considered as two organic inhibitors, the ratio of resin content to asphaltene content in the inhibitor should be significant to be effective [7].

3.5.5 Ionic liquids

Ionic liquids have local charge densities, which can disrupt the asphaltene self-association. The ionic liquid inhibitory strength increases when its anion has a high charge density (small size and high charge) and/or its cation has a low charge density (big size and low charge), which causes an acid-base interaction between the asphaltene and ionic liquid pair [9, 113]. Moreover, additional functional groups can improve the ionic liquid inhibitory effect. For instance, the addition of boronic acid enhances the efficiency of ionic inhibitors, although it is a weak acid [109]. Hu and Guo [9] found that ionic liquids with short alkyl tails inhibit asphaltene deposition better; on the contrary, Boukherissa et al. [109] and Subramanian et al. [113] demonstrated that ionic liquids with longer alkyl chains have a high inhibitory effect. If the ionic inhibitors are not chosen properly, they may even worsen asphaltene damage as a result of the ionic liquid self-aggregation [113].

3.6 Importance/application of inhibition

Based on the thermodynamic condition and crude oil composition, asphaltene deposition can happen in oil reservoirs, near wellbores, and along wells, pipelines, and surface facilities. Prevention of asphaltene precipitation is favored over removing its deposition because the preventive methods last for a longer period and reduce the number of production interruptions, leading to lower costs and risks in the oil industry. The compatibility, type, and concentration of selected inhibitors are the most critical parameters in designing asphaltene inhibitors; improper inhibitor design may cause a more severe deposition problem in comparison to an untreated condition. Moreover, the development of chemical inhibitors should still incorporate considerations on environmental, safety, and health impacts.

3.7 Costs of chemical inhibitors

In the previous sections, the technical aspect of different types of inhibitors are presented; however, the economic perspective should also be considered in screening the inhibitors for large-scale applications. In this section, the costs of common inhibitors are listed in Table 3.6. These cost numbers are obtained from major suppliers of the chemicals and we have

Table 3.6 The cost and required concentration of chemical inhibitors in a specific purity and quantity.

Inhibitor	Price (CAD$)	US$	Purity (%) (size for nanoparticles)	Quantity (kg or L)	Required concentration	Source
Surfactants						
Dodecylbenzenesulfonic acid	226	153	>95	1 L	0.0025%–16%	Aldrich
Linear alkylbenzenesulfonic acid	1220	1000	>96	1000 kg	0.1–2 wt%	Alibaba
Dioctyl sodium sulfosuccinate	495	317	>97	1 kg	20%	Aldrich
Triethanolamine lauryl ether sulfate	10	8	39%	1 kg	0.1%–2%	Alibaba
Sodium lauryl ether sulfate	1290	1000	70%	1000 kg	0.1%–2%	Alibaba
Benzoic acid	487	379	>99.5	3 kg	0.025% and 10%	Aldrich
Salicylic acid	279	170	>99	0.5 kg	0%–0.1%	Aldrich
Phthalic acid	263	179	>99.5	1 kg	0%–0.1%	Aldrich
Linoleic acid	184	131	58–74	1 L	0%–5%	Aldrich
Sodium octylbenzenesulfonate	155	129	97	0.005 kg	0%–4%	Aldrich
Sodium dodecylbenzenesulfonate	115	77.8	Technical grade	1 kg	0%–4%	Aldrich
Sodium dodecyl sulfate (SDS)	500	376	>98.5	1 kg	0.03%	Aldrich
Sebacic acid	111	81.8	>95	1 kg	Not mention	Aldrich
Caprylic acid	345	255	>98	10 kg	0%–5%	Aldrich
Lauric acid	285	211	>98	10 kg	0%–0.2%	Aldrich
Palmitic acid	701	522	>99	0.25 kg	0%–5%	Aldrich
Stearic acid	152	103	>95	5 kg	0.5%	Aldrich
Methylphenol	117	86	>98	1 kg	1%–20%	Aldrich

Continued

Table 3.6 The cost and required concentration of chemical inhibitors in a specific purity and quantity—cont'd

Inhibitor	Price (CAD$)	US$	Purity (%) (size for nanoparticles)	Quantity (kg or L)	Required concentration	Source
Ethylphenol	172	134	>97	1 kg	1%–16%	Aldrich
Butylphenol[a]	223.79	170	>96	0.025 kg	1%–12%	TCI Chemical
Hexylphenol	372	277	>98	0.025 kg	0%–6%	Aldrich
Octylphenol	200	138	99	0.005 kg	0%–6%	Aldrich
Nonylphenol	194	135	Technical grade	3 L	0%–24%	Aldrich
Dodecylphenol	74	57.2	Mixture of isomers	1 L	0%–16%	Aldrich
Renex 20	105.68	79.95	–	0.5 L	0–110 mg/L	Spectrum Chemical
Renex 40[a]	11	8.36	99	1 kg	0–110 mg/L	Alibaba
Renex 60[a]	2	1.52	99.9	1 kg	0%–16%	Alibaba
Renex 100[a]	3	2.28	>99	1 kg	0–110 mg/L	Alibaba
sec-Butylphenol	36.5	26.3	98	0.1 kg	4%	Aldrich
tert-Octylphenol	94.3	70.1	97	1 kg	8%	Aldrich
2-methoxy-4-methylphenol (Creosol)	65.8	51	>98	0.1 L	10%–50%	TCI Chemical
1-Octanol	239	185	>99	2 L	0.8–1.6	Aldrich
Dodecanol	153	146	98	3 kg	1%–25%	Aldrich
2-Ethyl-1-hexanol	94.5	70	>99	1 kg	1.25	Aldrich
Sorbitan monooleate (SM)	164	110	Span 80	1 L	5–10 vol%	Aldrich
Diethanoldodecylamine	356	276	>95	0.025 kg	5–10 vol%	TCI Chemical
Hexylamine	40.7	30.7	99%	0.1 L	1.25%	Aldrich
Dodecylamine	128	87.5	98	1 L	0.8%–1.6%	Aldrich
Hexadecylamine	65.5	53.5	90	0.5 kg	0.5%	Aldrich
tert-Butylphenol	67.8	55.8	99	1 kg	1%–16%	Aldrich

heptyl-Oxyphenol	101	89.6	97	0.005 kg	0%–6%	Aldrich
Triton X-100	86	71.9	Lab grade	1 L	0%–0.1%	Aldrich
Benzyl alcohol	367	338	>99	4 L	1.25%	Aldrich
Dodecylresorcinol	212	156	97	0.01 kg	0%–3%	Aldrich
Dodecyl-2-pyrrolidinone	87.5	74.7	99	0.25 L	5–10 vol%	Aldrich
Cetyl-Trimethylammonium bromide	385	313	>98	1 kg	0%–2%	Aldrich
Dodecyltrimethylammonium bromide	411	280	>98	0.1 kg	2%	Aldrich
Acetic acid	61.6	62.6	>99	1 L		Aldrich
Oleic acid	57.6	51.1	90	4 L	10%	Aldrich
Phenol	267	234	>99.5	0.5 kg	10%	Aldrich
Methanol	65.1	60.3	>99.9	1 L		Aldrich
isopropanol	45.8	62.6	70	1 L		Aldrich
Butylamine	69.8	53	99.5	1 L	0–100 vol%	Aldrich
Diethylenetriamine	63.8	62.3	99	1 L	5 vol%	Aldrich
Diethylaminopropylamine	234	177	>99	2 L	5 vol%	Aldrich
Cyclohexylamine	45.4	31.9	>99.9	0.1 L	5 vol%	Aldrich
Monoethanolamine	74.6	67.1	>98	1 L	5 vol%	Aldrich
Diethylamine	108	74.2	>99.5	1 L	5 vol%	Aldrich
Dimethylaminopropylamine	60.3	40.4	>98	1 L	5 vol%	Aldrich
Morpholine	173	148	>99	2.5 L	5 vol%	Aldrich
Polymers						
poly (Maleic anhydride l-octadecene)	153	147	–	1 kg	0%–0.1%	Aldrich

Continued

Table 3.6 The cost and required concentration of chemical inhibitors in a specific purity and quantity—cont'd

Inhibitor	Price (CAD$)	US$	Purity (%) (size for nanoparticles)	Quantity (kg or L)	Required concentration	Source
Nanoparticles						
Aluminum[a]	85.57	65	800 nm	0.1 kg	0–3 g/L	US Research Nanomaterials Inc
Alumina	105	79.3	<50 nm	0.05 kg	0–3 g/L	Aldrich
γ-Alumina[a]	130.32	99	80 nm	0.1 kg	0–3 g/L	US Research Nanomaterials Inc
Silica gel	402	280	Pore size = 60 A & 30–60 mesh particle size	1 kg	0–1 g/L	Supelco
Silica[a]	85.57	65	60–70 nm	0.1 kg	0.05–2 g/L	US Research Nanomaterials Inc
MgO	392	313	<50 nm	0.025 kg	0–3 g/L	Aldrich
TiO_2	121	108	<100 nm	0.005 kg	0–2.5 g/L	Aldrich
Fe_3O_4[a]	117.16	89	20–30 nm	0.1 kg	0%–10%	US Research Nanomaterials Inc
Fe_2O_3	42	33.5	<50 nm	0.005 kg	0–3 g/L	Aldrich
Co_3O_4	477	382	<50 nm	0.25 kg	0–2 g/L	Aldrich
Iron	195.65	148	800 nm	0.1 kg	0–3 g/L	US Research Nanomaterials Inc
Fumed silica gel	79.6	67.2	200–300 nm	0.5 kg	0–1 g/L	Supelco
ZrO_2	139	117	<100 nm	0.025 kg	1–3 g/L	Aldrich
NiO	390	313	<50 nm	0.25 kg	0–10 g/L	Aldrich
WO_3	207	156	<100 nm	0.025 kg	0%–3%	Aldrich
CaO	573	407	<160 nm	0.1 kg	0%–3%	Aldrich

Solvents							
Toluene	691	618	99.8		20 L	0%–50%	Aldrich
Nonylbenzene	574	445	97		0.25 kg	0%–6%	Alfa
Naphthalene	65	46.6	99		1 kg	0%–30%	Aldrich
Quinoline	130	96	>97		1 kg	15%	Aldrich
Indole	241	178	>99		1 kg	20%	Aldrich
Benzene	787	765	>99.9		16 L	0.1%	Aldrich
Xylene	514	685	>99		8 L	0%–30%	Aldrich
1,2,4-Trimethylbenzene	127	112	98		2 L	15%	Aldrich
Decalin	125	109	98		1 L	15%–20%	Aldrich
Tetralin	263	270	>97		4 L	15%–20%	Aldrich
Phenanthrene	441	394	98		0.5 kg	0%–20%	Aldrich
Pyridine	270	182	>99.9		1 L	15%	Aldrich
Benzothiophene	396	243	98		0.1 kg	15%–25%	Aldrich
Cyclohexane	105	101	99.5		1 L		Aldrich
Dicyclopentadiene	85.9	67.2			1 kg		Aldrich
Ionic liquids							
Butylpyridinium chloride	94.5	90.9	>98		0.025 kg	0%–4%	Aldrich
Octylpyridinium chloride[a]	787.21	598	98		0.1 kg	0%–4%	BOC science
Dodecylpyridinium chloride	423	328	>98		0.5 kg	0%–4%	TCI
1-Butyl-3-methylimidazolium bromide	114	105	>97		0.05 kg	0–4 g/L	Aldrich
1-Butyl-3-methylimidazolium chloride	428	399	>98		0.25 kg	0–4 g/L	Aldrich

Continued

Table 3.6 The cost and required concentration of chemical inhibitors in a specific purity and quantity—cont'd

Inhibitor	Price (CAD$)	US$	Purity (%) (size for nanoparticles)	Quantity (kg or L)	Required concentration	Source
1-Butyl-3-methylimidazolium thiocyanate	4360	4070	>95	1 kg	0.001 g/L	Aldrich
1-Butyl-3-methylimidazolium tetrafluoroborate	1200	911	>98	1 kg	0.001 g/L	Aldrich
1-Butyl-3-methylimidazolium nitrate	1220	758	>95	0.05 kg	0–10 Ionic to asphaltene ratio	Aldrich
1-Ethyl-3-methylimidazolium chloride	471	429	>95	1 kg	0.001–0.005 g/L	Aldrich
1-Butyl-3-methylimidazolium chloride	428	399	>98	0.25 kg	0.001–0.005 g/L	Aldrich
1-Hexyl-3-methylimidazolium chloride	187	164	>97	0.05 kg	0.001–0.005 g/L	Aldrich
1-Octyl-3-methylimidazolium chloride	82.3	63	98	0.025 kg	0.001–0.005 g/L	J&K Scientific
1-Decyl-3-methylimidazolium chloride	351	292	96	0.05 kg	0.001–0.005 g/L	Aldrich

[a]Either the CAD or the US$ value has been converted based on the Currency Exchange rate reported by Bank of Canada Oct 27, 2020.

chosen their prices based on the largest quantity available. These prices are expected to be lower at a larger scale purchase. Lab-synthesized inhibitors and organic inhibitors that were produced from particular crude oils were not available in the market; thus, it was impossible to determine a specific cost for them. Moreover, an approximation of the required concentration of inhibitors is introduced based on the references. Based on this information, there might be an expensive inhibitor that would be effective in a low dosage or a cheap one that needs a higher dosage to be efficient. Hence, Table 3.6 provides researchers and engineers with a summarized source in which the selection of a reasonable inhibitor can be made from an economical viewpoint. A rough economic analysis and screening of well-known inhibitors are also provided in the following chapters.

3.8 Final remarks

Asphaltene inhibitors are additives that change the fluid composition, and/or molecular coordination of the reservoir fluids to either inhibit or mitigate (delay) the asphaltene aggregation, precipitation, and deposition. Extensive research studies exist in this field, for which many inhibitors are available with two main categories of kinetic inhibitors and thermodynamic inhibitors. Kinetic inhibitors, such as surfactants, keep the asphaltene particles suspended and separated by attaching to them and creating steric hindrance around them. Thermodynamic inhibitors blend with crude oil and disrupt the ratio of saturate, aromatic, resin, and asphaltene in the crude oil; by changing the oil composition, the thermodynamic inhibitors mitigate the precipitation and deposition of asphaltenes, similar to the role of resins. For the kinetic inhibitors, properties such as chemical structure, polarity, acidity, basicity, and the alkyl tail structure and length control the inhibitory efficiency for a specific oil. For the thermodynamic inhibitors, the SARA fraction, acidity, and compatibility of the inhibitor with the problematic crude are the controlling parameters. The wide variations of inhibitor types in the market provide engineers with flexibility in designing a more efficient asphaltene deposition inhibition (or mitigation); although the variations will make the screening process complex to consider all aspects. Inhibitors have advantages compared to conventional treatment methods. For instance, upon using them, the fewer number of treatments and less production interruptions are required. They also cause fewer environmental side effects and increase the human safety and health benefits.

References

[1] M.J. Jeffries-Harris, C.P. Coppel, Solvent stimulation in low gravity oil reservoirs, J. Petrol. Tech. 21 (02) (1969) 167–175.
[2] G.D. Sutton, Composition and process for the removal of asphaltenic containing organic deposits from surfaces, Google Patents, 1975.
[3] M.E. Newberry, K.M. Barker, Method for the removal of asphaltenic deposits, Google Patents, 1983.
[4] L.C.R. Junior, M.S. Ferreira, A.C. da Silva Ramos, Inhibition of asphaltene precipitation in Brazilian crude oils using new oil soluble amphiphiles, J. Petrol. Sci. Eng. 51 (1–2) (2006) 26–36.
[5] H.H. Ibrahim, R.O. Idem, Interrelationships between asphaltene precipitation inhibitor effectiveness, asphaltenes characteristics, and precipitation behavior during n-heptane (light paraffin hydrocarbon)-induced asphaltene precipitation, Energy Fuel 18 (4) (2004) 1038–1048.
[6] H.H. Ibrahim, R.O. Idem, CO_2-miscible flooding for three Saskatchewan crude oils: interrelationships between asphaltene precipitation inhibitor effectiveness, asphaltenes characteristics, and precipitation behavior, Energy Fuel 18 (3) (2004) 743–754.
[7] T.A. Al-Sahhaf, M.A. Fahim, A.S. Elkilani, Retardation of asphaltene precipitation by addition of toluene, resins, deasphalted oil and surfactants, Fluid Phase Equilib. 194 (2002) 1045–1057.
[8] L. Goual, A. Firoozabadi, Effect of resins and DBSA on asphaltene precipitation from petroleum fluids, AICHE J. 50 (2) (2004) 470–479.
[9] Y.-F. Hu, T.-M. Guo, Effect of the structures of ionic liquids and alkylbenzene-derived amphiphiles on the inhibition of asphaltene precipitation from CO_2-injected reservoir oils, Langmuir 21 (18) (2005) 8168–8174.
[10] E. Rogel, O. Leon, Y. Espidel, Y. Gonzalez, Asphaltene stability in crude oils, SPE Product. Facil. 16 (02) (2001) 84–88.
[11] O. León, E. Rogel, A. Urbina, A. Andújar, A. Lucas, Study of the adsorption of alkyl benzene-derived amphiphiles on asphaltene particles, Langmuir 15 (22) (1999) 7653–7657.
[12] D. Subramanian, A. Firoozabadi, Effect of surfactants and water on inhibition of asphaltene precipitation and deposition, in: Abu Dhabi International Petroleum Exhibition and Conference, Society of Petroleum Engineers, 2015.
[13] A. Naseri, M. Nikazar, S. Dehghani, B. Dabir, O. Gohari, The role of Inhibitors' molecular structure on Asphaltene deposition in reservoir conditions, Pet. Sci. Technol. 29 (10) (2011) 988–999.
[14] K. Kraiwattanawong, H.S. Fogler, S.G. Gharfeh, P. Singh, W.H. Thomason, S. Chavadej, Effect of asphaltene dispersants on aggregate size distribution and growth, Energy Fuel 23 (3) (2009) 1575–1582.
[15] S. Peramanu, P.F. Clarke, B.B. Pruden, Flow loop apparatus to study the effect of solvent, temperature and additives on asphaltene precipitation, J. Petrol. Sci. Eng. 23 (2) (1999) 133–143.
[16] P. Permsukarome, C. Chang, H.S. Fogler, Kinetic study of asphaltene dissolution in amphiphile/alkane solutions, Ind. Eng. Chem. Res. 36 (9) (1997) 3960–3967.
[17] S.M. Hashmi, A. Firoozabadi, Effective removal of asphaltene deposition in metal-capillary tubes, SPE J. 21 (05) (2016) 1747–1754.
[18] A.C. da Silva Ramos, L. Haraguchi, F.R. Notrispe, W. Loh, R.S. Mohamed, Interfacial and colloidal behavior of asphaltenes obtained from Brazilian crude oils, J. Petrol. Sci. Eng. 32 (2–4) (2001) 201–216.
[19] A. Al Sultan, M. Zirrahi, H. Hassanzadeh, J. Abedi, Effect of the surfactant on Asphaltene deposition on stainless-steel and glass surfaces, Energy Fuel 32 (4) (2018) 5635–5642.

[20] J. Wang, C. Li, L. Zhang, G. Que, Z. Li, The properties of asphaltenes and their interaction with amphiphiles, Energy Fuel 23 (7) (2009) 3625–3631.
[21] S.M. Hashmi, K.X. Zhong, A. Firoozabadi, Acid–base chemistry enables reversible colloid-to-solution transition of asphaltenes in non-polar systems, Soft Matter 8 (33) (2012) 8778–8785.
[22] T.S. Loureiro, L.C.M. Palermo, L.S. Spinelli, Influence of precipitation conditions (n-heptane or carbon dioxide gas) on the performance of asphaltene stabilizers, J. Petrol. Sci. Eng. 127 (2015) 109–114.
[23] L.Z. Pillon, Effect of dispersants and flocculants on the colloidal stability of asphaltenes, Pet. Sci. Technol. 19 (7–8) (2001) 863–873.
[24] F. Marcano, L.G. Moura, F.M. Cardoso, P.T. Rosa, Evaluation of the chemical additive effect on asphaltene aggregation in dead oils: a comparative study between ultraviolet–visible and near-infrared-laser light scattering techniques, Energy Fuel 29 (5) (2015) 2813–2822.
[25] P.F. Clarke, B.B. Pruden, Asphaltene precipitation from Cold Lake and Athabasca bitumens, Pet. Sci. Technol. 16 (3–4) (1998) 287–305.
[26] C.-L. Chang, H. Fogler, Asphaltene stabilization in alkyl solvents using oil-soluble amphiphiles, in: SPE International Symposium on Oilfield Chemistry, Society of Petroleum Engineers, 1993.
[27] E. Zahedi, M. Vafaie-Sefti, M. Shadman, H. Naderi, M. Amiri, A. Noorbakhsh, Experimental investigation of dodecylbenzene sulfonic acid and toluene dispersants on asphaltene precipitation of dead and live oil, Pet. Sci. Technol. 35 (7) (2017) 653–660.
[28] I. Auflem, T. Havre, J. Sjöblom, Near-IR study on the dispersive effects of amphiphiles and naphthenic acids on asphaltenes in model heptane-toluene mixtures, Colloid Polym. Sci. 280 (8) (2002) 695–700.
[29] C.-L. Chang, H.S. Fogler, Stabilization of asphaltenes in aliphatic solvents using alkylbenzene-derived amphiphiles. 1. Effect of the chemical structure of amphiphiles on asphaltene stabilization, Langmuir 10 (6) (1994) 1749–1757.
[30] E. Rogel, E. Contreras, O. Leon, An experimental theoretical approach to the activity of amphiphiles as asphaltene stabilizers, Pet. Sci. Technol. 20 (7–8) (2002) 725–739.
[31] M.M. Shadman, M. Dehghanizadeh, A.H. Saeedi Dehaghani, M. Vafaie Sefti, N. Mokhtarian, An investigation of the effect of aromatic, anionic and nonionic inhibitors on the onset of asphaltene precipitation, J. Oil Gas Petrochem. Technol. 1 (2014) 17–28.
[32] C.R. Mansur, A.R. de Melo, E.F. Lucas, Determination of asphaltene particle size: influence of flocculant, additive, and temperature, Energy Fuel 26 (8) (2012) 4988–4994.
[33] A.H.S. Dehaghani, M.H. Badizad, Inhibiting asphaltene precipitation from Iranian crude oil using various dispersants: experimental investigation through viscometry and thermodynamic modelling, Fluid Phase Equilib. 442 (2017) 104–118.
[34] G. Piro, L.B. Canonico, G. Galbariggi, L. Bertero, C. Carniani, Asphaltene adsorption onto formation rock: an approach to asphaltene formation damage prevention, SPE Product. Facil. 11 (03) (1996) 156–160.
[35] R.S. Mohamed, W. Loh, A.C. Ramos, C.C. Delgado, V.R. Almeida, Reversibility and inhibition of asphaltene precipitation in Brazilian crude oils, Pet. Sci. Technol. 17 (7–8) (1999) 877–896.
[36] M. Madhi, R. Kharrat, T. Hamoule, Screening of inhibitors for remediation of asphaltene deposits: Experimental and modeling study, Petroleum 4 (2) (2018) 168–177.
[37] S.M. Hashmi, A. Firoozabadi, Tuning size and electrostatics in non-polar colloidal asphaltene suspensions by polymeric adsorption, Soft Matter 7 (18) (2011) 8384–8391.

[38] S.M. Hashmi, L.A. Quintiliano, A. Firoozabadi, Polymeric dispersants delay sedimentation in colloidal asphaltene suspensions, Langmuir 26 (11) (2010) 8021–8029.

[39] I.A. Wiehe, T.G. Jermansen, Design of synthetic dispersants for asphaltenes, Pet. Sci. Technol. 21 (3–4) (2003) 527–536.

[40] S. Kashefi, A. Shahrabadi, S. Jahangiri, M.N. Lotfollahi, H. Bagherzadeh, Investigation of the performance of several chemical additives on inhibition of asphaltene precipitation, Energy Source A: Recov. Utilizat. Environ. Effect 38 (24) (2016) 3647–3652.

[41] H. Lian, J.-R. Lin, T.F. Yen, Peptization studies of asphaltene and solubility parameter spectra, Fuel 73 (3) (1994) 423–428.

[42] M.A. Karambeigi, M. Nikazar, R. Kharrat, Experimental evaluation of asphaltene inhibitors selection for standard and reservoir conditions, J. Petrol. Sci. Eng. 137 (2016) 74–86.

[43] L.A. Alcázar-Vara, L.S. Zamudio-Rivera, E. Buenrostro-González, R.L. Hernández-Altamirano, V.Y. Mena-Cervantes, J.F. Ramírez-Pérez, Multifunctional properties of Zwitterionic liquids. Application in enhanced oil recovery and Asphaltene aggregation phenomena, Ind. Eng. Chem. Res. 54 (11) (2015) 2868–2878.

[44] E. Joonaki, R. Burgass, B. Tohidi, Experimental and modelling study on application of new class of asphaltene inhibitors for enhanced oil recovery purposes: adsorption and wettability alteration, in: Abu Dhabi International Petroleum Exhibition and Conference, Society of Petroleum Engineers, 2016.

[45] Y.-J. Lin, et al., Characterizing Asphaltene deposition in the presence of chemical dispersants in porous media micromodels, Energy Fuel 31 (11) (2017) 11660–11668.

[46] L. Goual, M. Sedghi, X. Wang, Z. Zhu, Asphaltene aggregation and impact of alkylphenols, Langmuir 30 (19) (2014) 5394–5403.

[47] O. León, E. Contreras, E. Rogel, Amphiphile adsorption on asphaltene particles: adsorption isotherms and asphaltene stabilization, Colloids Surf. A Physicochem. Eng. Asp. 189 (1–3) (2001) 123–130.

[48] O. Leon, E. Contreras, E. Rogel, G. Dambakli, J. Espidel, S. Acevedo, The influence of the adsorption of amphiphiles and resins in controlling asphaltene flocculation, Energy Fuel 15 (5) (2001) 1028–1032.

[49] G. Gonzalez, A. Middea, Peptization of asphaltene by various oil soluble amphiphiles, Colloids Surf. 52 (1991) 207–217.

[50] M. Barcenas, P. Orea, Molar-mass distributions of asphaltenes in the presence of inhibitors: experimental and computer calculations, Energy Fuel 25 (5) (2011) 2100–2108.

[51] M. Yakubov, S. Yakubova, D. Borisov, G. Romanov, K. Yakubson, Asphaltene precipitation inhibitors and phase behavior control for bitumen recovery by solvent injection, in: SPE Heavy Oil Conference-Canada, Society of Petroleum Engineers, 2014.

[52] D. Merino-Garcia, S.I. Andersen, Interaction of asphaltenes with nonylphenol by microcalorimetry, Langmuir 20 (4) (2004) 1473–1480.

[53] M. Barcenas, P. Orea, E. Buenrostro-González, L.S. Zamudio-Rivera, Y. Duda, Study of medium effect on asphaltene agglomeration inhibitor efficiency, Energy Fuel 22 (3) (2008) 1917–1922.

[54] C.-L. Chang, H. Scott Fogler, Peptization and coagulation of asphaltenes in apolar media using oil-soluble polymers, Fuel Sci. Technol. Int. 14 (1–2) (1996) 75–100.

[55] P.F. Clarke, B.B. Pruden, Asphaltene precipitation: detection using heat transfer analysis, and inhibition using chemical additives, Fuel 76 (7) (1997) 607–614.

[56] J. Pereira, J. Delgado-Linares, A. Briones, M. Guevara, C. Scorzza, J.-L. Salager, The effect of solvent nature and dispersant performance on asphaltene precipitation from diluted solutions of instable crude oil, Pet. Sci. Technol. 29 (23) (2011) 2432–2440.

[57] K. Safaie, A.R.S. Nazar, Evaluation of asphaltene inhibitors effect on aggregation coupled sedimentation process, J. Dispers. Sci. Technol. 35 (3) (2014) 329–337.

[58] T.E. Chávez-Miyauchi, L.S. Zamudio-Rivera, V. Barba-López, E. Buenrostro-Gonzalez, J.M. Martínez-Magadán, N-aryl amino-alcohols as stabilizers of asphaltenes, Fuel 110 (2013) 302–309.
[59] J.J. Adams, Asphaltene adsorption, a literature review, Energy Fuel 28 (5) (2014) 2831–2856.
[60] F. Shojaati, M. Riazi, S.H. Mousavi, Z. Derikvand, Experimental investigation of the inhibitory behavior of metal oxides nanoparticles on asphaltene precipitation, Colloids Surf. A Physicochem. Eng. Asp. 531 (2017) 99–110.
[61] N.N. Nassar, A. Hassan, P. Pereira-Almao, Effect of surface acidity and basicity of aluminas on asphaltene adsorption and oxidation, J. Colloid Interface Sci. 360 (1) (2011) 233–238.
[62] G. Fritschy, E. Papirer, Interactions between a bitumen, its components and model fillers, Fuel 57 (11) (1978) 701–704.
[63] H. Alboudwarej, D. Pole, W.Y. Svrcek, H.W. Yarranton, Adsorption of asphaltenes on metals, Ind. Eng. Chem. Res. 44 (15) (2005) 5585–5592.
[64] D. Dudášová, S. Simon, P.V. Hemmingsen, J. Sjöblom, Study of asphaltenes adsorption onto different minerals and clays: Part 1. Experimental adsorption with UV depletion detection, Colloids Surf. A Physicochem. Eng. Asp. 317 (1–3) (2008) 1–9.
[65] N. Hosseinpour, A.A. Khodadadi, A. Bahramian, Y. Mortazavi, Asphaltene adsorption onto acidic/basic metal oxide nanoparticles toward in situ upgrading of reservoir oils by nanotechnology, Langmuir 29 (46) (2013) 14135–14146.
[66] N.N. Nassar, A. Hassan, P. Pereira-Almao, Metal oxide nanoparticles for asphaltene adsorption and oxidation, Energy Fuel 25 (3) (2011) 1017–1023.
[67] C.A. Franco, N.N. Nassar, M.A. Ruiz, P. Pereira-Almao, F.B. Cortés, Nanoparticles for inhibition of asphaltenes damage: adsorption study and displacement test on porous media, Energy Fuel 27 (6) (2013) 2899–2907.
[68] S. Betancur, J.C. Carmona, N.N. Nassar, C.A. Franco, F.B. Cortés, Role of particle size and surface acidity of silica gel nanoparticles in inhibition of formation damage by asphaltene in oil reservoirs, Ind. Eng. Chem. Res. 55 (21) (2016) 6122–6132.
[69] M. Madhi, A. Bemani, A. Daryasafar, M.R. Khosravi Nikou, Experimental and modeling studies of the effects of different nanoparticles on asphaltene adsorption, Pet. Sci. Technol. 35 (3) (2017) 242–248.
[70] M. Mohammadi, M. Akbari, Z. Fakhroueian, A. Bahramian, R. Azin, S. Arya, Inhibition of asphaltene precipitation by TiO_2, SiO_2, and ZrO_2 nanofluids, Energy Fuel 25 (7) (2011) 3150–3156.
[71] M. Igder, N. Hosseinpour, A. Amrollahi Biyouki, A. Bahramian, Control of asphaltene aggregation in reservoir model oils along the production streamline by Fe_3O_4 and NiO nanoparticles, Energy Fuel 32 (6) (2018) 6689–6697.
[72] Y. Kazemzadeh, M. Malayeri, M. Riazi, R. Parsaei, Impact of Fe_3O_4 nanoparticles on asphaltene precipitation during CO_2 injection, J. Nat. Gas Sci. Eng. 22 (2015) 227–234.
[73] N.N. Nassar, Asphaltene adsorption onto alumina nanoparticles: kinetics and thermodynamic studies, Energy Fuel 24 (8) (2010) 4116–4122.
[74] T. Lu, Z. Li, W. Fan, X. Zhang, Q. Lv, Nanoparticles for inhibition of asphaltenes deposition during CO_2 flooding, Ind. Eng. Chem. Res. 55 (23) (2016) 6723–6733.
[75] Z. Romero, et al., Application and evaluation of a nanofluid containing nanoparticles for asphaltenes inhibition in well CPSXL4, in: OTC Brasil, 2013. Offshore Technology Conference.
[76] C. Carpenter, Application of a Nanofluid for asphaltene inhibition in Colombia, J. Petrol. Tech. 66 (02) (2014) 117–119.
[77] N.N. Nassar, A. Hassan, P. Pereira-Almao, Effect of the particle size on asphaltene adsorption and catalytic oxidation onto alumina particles, Energy Fuel 25 (9) (2011) 3961–3965.

[78] S.I. Hashemi, B. Fazelabdolabadi, S. Moradi, A.M. Rashidi, A. Shahrabadi, H. Bagherzadeh, On the application of NiO nanoparticles to mitigate in situ asphaltene deposition in carbonate porous matrix, Appl. Nanosci. 6 (1) (2016) 71–81.
[79] N.N. Nassar, M.E. Al-Jabari, M.M. Husein, Removal of asphaltenes from heavy oil by nickel nano and micro particle adsorbents, in: Proceedings of the IASTED International Conference, vol. 615, 2008, p. 808.
[80] F.B. Cortés, J.M. Mejía, M.A. Ruiz, P. Benjumea, D.B. Riffel, Sorption of asphaltenes onto nanoparticles of nickel oxide supported on nanoparticulated silica gel, Energy Fuel 26 (3) (2012) 1725–1730.
[81] Y. Kazemzadeh, S.E. Eshraghi, K. Kazemi, S. Sourani, M. Mehrabi, Y. Ahmadi, Behavior of asphaltene adsorption onto the metal oxide nanoparticle surface and its effect on heavy oil recovery, Ind. Eng. Chem. Res. 54 (1) (2015) 233–239.
[82] C. Ovalles, E. Rogel, H. Morazan, K. Chen, M.E. Moir, The use of nonylphenol formaldehyde resins for preventing asphaltene precipitation in vacuum residues and hydroprocessed petroleum samples, Pet. Sci. Technol. 34 (4) (2016) 379–385.
[83] A.F. Lima, C.R. Mansur, E.F. Lucas, G. González, Polycardanol or sulfonated polystyrene as flocculants for asphaltene dispersions, Energy Fuel 24 (4) (2009) 2369–2375.
[84] P. Juyal, V. Ho, A. Yen, S.J. Allenson, Reversibility of asphaltene flocculation with chemicals, Energy Fuel 26 (5) (2012) 2631–2640.
[85] S.M. Hashmi, A. Firoozabadi, Effect of dispersant on asphaltene suspension dynamics: aggregation and sedimentation, J. Phys. Chem. B 114 (48) (2010) 15780–15788.
[86] S.M. Hashmi, A. Firoozabadi, Controlling nonpolar colloidal asphaltene aggregation by electrostatic repulsion, Energy Fuel 26 (7) (2012) 4438–4444.
[87] A.M.A. Ghaffar, K.I. Kabel, R.K. Farag, N. Maysour, M.A. Zahran, Synthesis of poly (dodecyl phenol formaldehyde)-b-poly (oxypropylene) block copolymer, and evaluation as asphaltene inhibitor and dispersant, Res. Chem. Intermed. 41 (1) (2015) 443–455.
[88] T.S.E. Chávez-Miyauchi, L.S. Zamudio-Rivera, V. Barba-López, Aromatic polyisobutylene succinimides as viscosity reducers with asphaltene dispersion capability for heavy and extra-heavy crude oils, Energy Fuel 27 (4) (2013) 1994–2001.
[89] M. Alhreez, D. Wen, L. Ali, A novel inhibitor for controlling Iraqi asphaltene problems, in: Environmental Impacts of the Oil and Gas Industries: Kurdistan Region of Iraq as a Case Study (EIOGI), 2017 International Conference on, IEEE, 2017, pp. 37–41.
[90] H. Firoozinia, K.F.H. Abad, A. Varamesh, A comprehensive experimental evaluation of asphaltene dispersants for injection under reservoir conditions, Pet. Sci. 13 (2) (2016) 280–291.
[91] M. Bouts, R. Wiersma, H. Muijs, A. Samuel, An evaluation of new aspaltene inhibitors; laboratory study and field testing, J. Petrol. Tech. 47 (09) (1995) 782–787.
[92] D.F. Smith, G.C. Klein, A.T. Yen, M.P. Squicciarini, R.P. Rodgers, A.G. Marshall, Crude oil polar chemical composition derived from FT− ICR mass spectrometry accounts for asphaltene inhibitor specificity, Energy Fuel 22 (5) (2008) 3112–3117.
[93] C. Chen, J. Guo, N. An, Y. Pan, Y. Li, Q. Jiang, Study of asphaltene dispersion and removal for high-asphaltene oil wells, Pet. Sci. 9 (4) (2012) 551–557.
[94] S. Kashefi, A. Shahrabadi, M.N. Lotfollahi, A. Varamesh, A new polymeric additive as asphaltene deposition inhibitor in CO_2 core flooding, Korean J. Chem. Eng. 33 (11) (2016) 3273–3280.
[95] M. Wang, J. Kaufman, X. Chen, C. Sungail, Development and evaluation of nonionic polymeric surfactants as asphaltene inhibitors, in: SPE International Symposium on Oilfield Chemistry, Society of Petroleum Engineers, 2015.
[96] A. Khormali, Asphaltene precipitation and inhibition in carbonate reservoirs, Pet. Sci. Technol. 35 (5) (2017) 515–521.

[97] A. Khormali, A.R. Sharifov, D.I. Torba, The control of asphaltene precipitation in oil wells, Pet. Sci. Technol. 36 (6) (2018) 443–449.
[98] S. Subramanian, S. Simon, J. Sjöblom, Interaction between asphaltenes and fatty-alkylamine inhibitor in bulk solution, J. Dispers. Sci. Technol. 39 (2) (2018) 163–173.
[99] S. Subramanian, L. Buscetti, S.B. Simon, M. Sacré, J. Sjöblom, Influence of fatty-Alkylamine Amphiphile on the Asphaltene adsorption/deposition at the solid/liquid Interface under precipitating conditions, Energy Fuel 32 (4) (2018) 4772–4782.
[100] S. Schantz, W. Stephenson, Asphaltene deposition: development and application of polymeric asphaltene dispersants, in: SPE Annual Technical Conference and Exhibition, Society of Petroleum Engineers, 1991.
[101] A. Aquino-Olivos, E. Buenrostro-Gonzalez, S.I. Andersen, C. Lira-Galeana, Investigations of inhibition of asphaltene precipitation at high pressure using bottomhole samples, Energy Fuel 15 (1) (2001) 236–240.
[102] J. Marugán, J.A. Calles, J. Dufour, R. Giménez-Aguirre, J.L. Peña, D. Merino-García, Characterization of the asphaltene onset region by focused-beam laser reflectance: a tool for additives screening, Energy Fuel 23 (3) (2008) 1155–1161.
[103] A. Albannay, T. Masuzawa, C.A. Dabbouk, Alternative methods in preventing asphaltene deposition, in: Abu Dhabi International Petroleum Exhibition and Conference, Society of Petroleum Engineers, 2010.
[104] H. Yonebayashi, Case study of Asphaltene flow assurance risk mitigation-emerging numerical model to evaluate inhibitor efficiency, in: 78th EAGE Conference and Exhibition 2016, 2016.
[105] H. Yonebayashi, X. Zhang, Emerging numerical modelling technique to evaluate asphaltene inhibitor efficiency during entire field life, in: SPE Asia Pacific Oil & Gas Conference and Exhibition, Society of Petroleum Engineers, 2016.
[106] P.M. Spiecker, K.L. Gawrys, C.B. Trail, P.K. Kilpatrick, Effects of petroleum resins on asphaltene aggregation and water-in-oil emulsion formation, Colloids Surf. A Physicochem. Eng. Asp. 220 (1–3) (2003) 9–27.
[107] N. Carnahan, J.L. Salager, R. Anton, Effect of resins on stability of asphaltenes, in: Offshore Technology Conference, Offshore Technology Conference, 2007.
[108] E.F. Ghloum, M. Al-Qahtani, A. Al-Rashid, Effect of inhibitors on asphaltene precipitation for Marrat Kuwaiti reservoirs, J. Petrol. Sci. Eng. 70 (1) (2010) 99–106.
[109] M. Boukherissa, F. Mutelet, A. Modarressi, A. Dicko, D. Dafri, M. Rogalski, Ionic liquids as dispersants of petroleum asphaltenes, Energy Fuel 23 (5) (2009) 2557–2564.
[110] L.F.B. Moreira, E.F. Lucas, G. González, Stabilization of asphaltenes by phenolic compounds extracted from cashew-nut shell liquid, J. Appl. Polym. Sci. 73 (1) (1999) 29–34.
[111] E. Monjezi, A. Rostami, E. Joonaki, S.R. Shadizadeh, New application of henna extract as an asphaltene inhibitor: an experimental study, Asia Pac. J. Chem. Eng. 11 (6) (2016) 1027–1034.
[112] A.S. Ogunlaja, E. Hosten, Z.R. Tshentu, Dispersion of asphaltenes in petroleum with ionic liquids: evaluation of molecular interactions in the binary mixture, Ind. Eng. Chem. Res. 53 (48) (2014) 18390–18401.
[113] D. Subramanian, K. Wu, A. Firoozabadi, Ionic liquids as viscosity modifiers for heavy and extra-heavy crude oils, Fuel 143 (2015) 519–526.
[114] A.M. Atta, A.O. Ezzat, M.M. Abdullah, A.I. Hashem, Effect of different families of hydrophobic anions of imadazolium ionic liquids on asphaltene dispersants in heavy crude oil, Energy Fuel 31 (8) (2017) 8045–8053.
[115] M. Mohammadi, M. Dadvar, B. Dabir, TiO_2/SiO_2 nanofluids as novel inhibitors for the stability of asphaltene particles in crude oil: mechanistic understanding, screening, modeling, and optimization, J. Mol. Liq. 238 (2017) 326–340.

CHAPTER FOUR

Properties of asphaltene chemical inhibitors

4.1 Characterization techniques to determine the properties of chemical inhibitors

4.1.1 Chemical structure

Chemical structure controls potential bonding forces between the asphaltene and chemical inhibitors. Fourier transform infrared (FTIR) spectroscopy reveals the chemical structure by exciting the molecules and adsorbing the emitted energy. The results from FTIR are recorded in the infrared spectrum, which show different bonds at specific peaks. For instance, aromatics and carbon double bonds have an absorption peak at 1600 cm^{-1}, and free N—H groups are distinguished by a peak at 3480 cm^{-1} [1]. In the research studies, FTIR is employed to analyze surfactants [1], polymers [2], and organic inhibitors [3]. In addition, FTIR is used to reveal the inhibition mechanisms by determining the changes in the frequency of bonding; for instance, it is used to evaluate the interactions of the polar groups in chemical inhibitors such as surfactant and ionic liquids, with asphaltene [2, 4, 5]. FTIR is commonly used to identify the composition of synthesized nanoparticles. In addition, X-ray diffraction (XRD) analysis can be used to measure the percentage of each component in the inhibitor [6]. Hydrogen and carbon nuclear magnetic resonance spectroscopy (H-NMR and C-NMR, respectively) are other analytical tools that provide information about the chemical structure of the synthetized inhibitors (such as ionic liquids) [2, 4, 5].

4.1.2 Chemical molecular weight

Molecular weight is considered as an influencing parameter for polymers used as asphaltene inhibitors. Gel permeation chromatography (GPC) and vapor pressure osmometry (VPO) techniques are two measurement methods that provide information about the molecular weight; they are commonly

Asphaltene Deposition Control by Chemical Inhibitors
https://doi.org/10.1016/B978-0-323-90510-7.00003-3

used to calculate the molecular weight of the asphaltenes, polymers, and organic liquids [7, 8].

4.1.2.1 Vapor pressure osmometry (VPO)

Vapor pressure osmometry (VPO) measurement is based on the vapor pressure difference due to the addition of low concentration additives (to a solvent) compared to a pure solvent. To obtain the vapor pressure difference, the vapor pressures are measured in two thermistors, which are surrounded with the vapor of pure solvent. One of the thermistors carries the droplet of pure solvent and the other one has a droplet of solvent-additive. Based on the molar mass (M), the voltage would change at different temperatures which can be calculated by the following equation [10]:

$$\Delta V = \frac{KC_2}{M} \tag{4.1}$$

where ΔV denotes the voltage difference between two thermistors, C_2 is the target fluid concentration, and K shows the calibration constant [10]. Fig. 4.1 depicts the schematic of vapor pressure osmometry device.

4.1.2.2 Gel permeation chromatography (GPC)

Gel permeation chromatography (GPC) relies on separating components in the solution according to their molecular size contrasts. In this method, the target component is dissolved in a solvent and is passed through a GPC

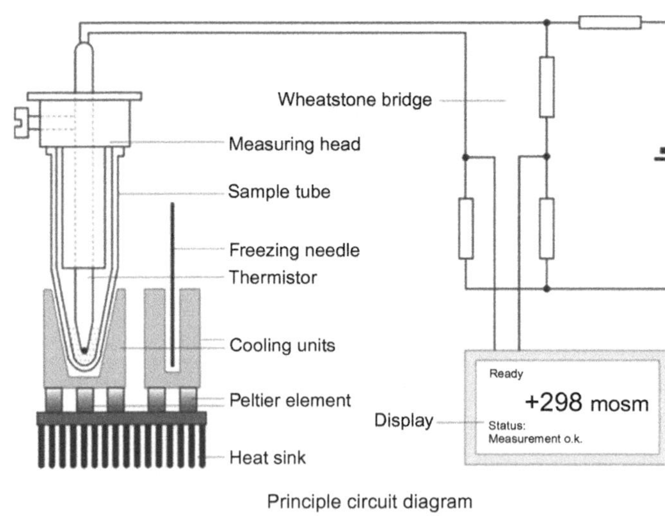

Fig. 4.1 Schematic representation of a vapor pressure osmometer [9].

column. The dissolved component will travel at different velocities based on its molecular size and molar mass. In this technique, the molar-mass average number (M_n) can be calculated using the following formula [7]:

$$M_n = \frac{\sum N_i M_i}{\sum N_i} \tag{4.2}$$

where N_i represents the number of moles and M_i is the molar mass of each component. Moreover, GPC provides molar mass distribution, measuring the distribution of the molecular mass for the asphaltene clusters [7]. However, it suffers from some limitations, including the underestimation of the molecular weight due to the lack of a calibration-specific standard, and overestimation due to molecular association and the adsorption of some compounds in the column [7]. Commonly, polystyrene is used for calibration, which may have a different chemical structure than the target asphaltene molecule.

4.1.3 SARA analysis

The SARA test identifies the amount of saturates, aromatics, resins, and asphaltenes in the crude oil and crude oil derivatives. The resins are natural asphaltene inhibitors. Therefore, a fluid with a high resin percentage has a higher potential to serve as an inhibitor. The SARA analysis can be used to assess the effectiveness of deasphalted oils, resins, and other processed hydrocarbons as inhibitors [8]. The procedure is initiated by separating the asphaltene part with an alkane solvent and adsorbing the residue (maltene) on silica gel. Normal alkanes, toluene, and toluene/methanol solvents are used to extract saturates, aromatics, and resins fraction, respectively, through chromatography [11].

4.1.4 Elemental analysis

Elemental analysis shows the concentration of hydrogen (H), carbon (C), oxygen (O), sulfur (S), and nitrogen (N) in fluids or solids, which is a useful source of information for organic inhibitors. For example, the percentage of H/C can indicate the acidity of the tested fluids or solid [8]. Based on the ASTM D5291 test [12], the sample material is combusted in the elemental analysis method at an elevated temperature in the presence of pure oxygen where carbon, hydrogen, and nitrogen are converted to carbon dioxide, water vapor, and elemental nitrogen, respectively. Subsequently, the amount of each gas is determined in the resultant gas stream. Sulfur oxides are removed with calcium oxide in the secondary combustion process. The rest of the gases is carried by helium over a hot copper train, NaOH, and magnesium perchlorate to reduce NO_x to N_2, to remove carbon dioxide, and to remove water vapor, respectively. Then, the remaining elemental nitrogen, carbon, and hydrogen are measured by a thermal conductivity cell and selective infrared cells [12].

4.1.5 Solubility parameter

The solubility of organic inhibitors in problematic oils is an important parameter that is used to examine the inhibitor suitability. The solubility of two fluids is defined by free Gibbs energy changes and the solubility parameter. The free energy change (ΔG_m) of a dissolution, defined by the following equation, should be negative:

$$\Delta G_m = \Delta H_m - T\Delta S_m \tag{4.3}$$

where ΔS_m denotes the entropy change, ΔH_m is the enthalpy change (mixing heat), and T represents the absolute temperature. Since ΔS_m is always positive, ΔH_m will be the determinative parameter and should be small or zero so that the inhibitor dissolves in a problematic oil.

The solubility parameter, δ, is defined as the square root of the energy change for complete isothermal vaporization of the saturated liquid, to form an ideal gas (ΔU^V), divided by the liquid molar volume (V^l), as introduced below [13]:

$$\delta = \sqrt{\frac{\Delta U^v}{V^l}} \tag{4.4}$$

The solubility parameter of unknown components can be determined by measuring their solubility in several solvents whose δ values are known [14]. According to the Hildebrand-Scatchard theory, ΔH_m is defined by the following equation [13]:

$$\Delta H_m = V_m(\delta_1 - \delta_2)^2\phi_1\phi_2 \tag{4.5}$$

where V_m is the total volume of the mixture, and ϕ and δ stand for the volume fraction and solubility parameter, respectively. The subscripts of 1 and 2 are designated to the solvent and solute, respectively. Consequently, the free energy change will be negative if $\delta_1 \simeq \delta_2$.

4.1.6 Charge density in ionic liquids

The charge density of the ions in ionic liquids is the main representative of the ionic liquid effectiveness for asphaltene inhibition and the charge density of ions is related to their charge and size. The ion charge is measurable with respect to the atomic charge, which is available in the periodic table. The ions' radii is also available in the literature [15–17]. Moreover, the variation of deionized water pH due to mixing with ionic liquids represents their acidity which can affect the ionic liquid inhibitory feature [18].

4.1.7 Nanoparticle characterization

4.1.7.1 Nanoparticle composition

In case of having nanocomposites, the percentage of each type of nanoparticles becomes important for the nanocomposite inhibitory effect. The energy-dispersive X-ray spectroscopy (EDX) is used to conduct the elemental analysis of nanocomposites. This test shows the concentration of each nanoparticle in the synthesized nanocomposites [6].

4.1.7.2 Nanoparticle shape

Although there is no discussion about the effect of nanoparticles' shape as an effective parameter for their inhibitory effect, this parameter is always reported as one of the main characteristics of nanoparticles. Nanoparticle shape may be recognized as an effective parameter for the shape of asphaltene aggregation; thus, it needs a systematic review. X-ray diffraction (XRD), scanning electron microscopy (SEM), and transmission electron microscopy (TEM) are three common tests that can reveal the shape of nanoparticles used for the asphaltene inhibition. XRD reports an intensity (count of diffracted angle per second) versus diffraction angle plot through which the shape of the nanoparticles is determined [19]. XRD is employed to measure the crystallinity and the analysis of phases in the nanoparticles and nanocomposites [6]. SEM and TEM show the shape of the nanoparticles by taking high magnification and resolution pictures from them. These pictures can reveal the size distribution and morphology of the nanoparticles (Fig. 4.2) [20].

4.1.7.3 Nanoparticle specific surface

The specific surface is an effective parameter for the nanoparticle's ability to adsorb asphaltene. The specific surface area of the nanoparticles can be measured through the Brunauer, Emmett, and Teller (BET) method. It determines the physical gas adsorption (usually nitrogen) on the solid surface and

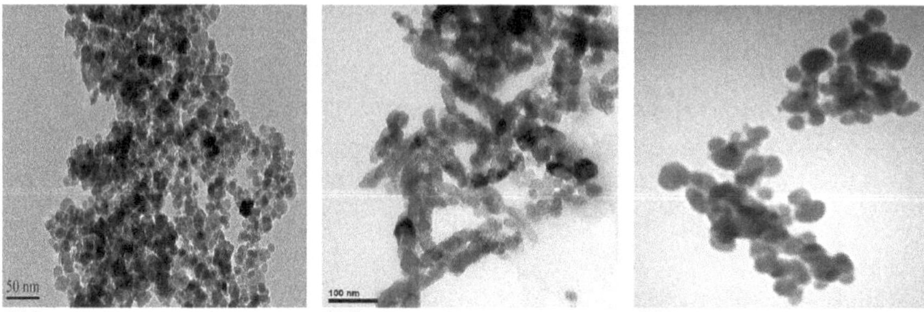

(A) (B) (C)

Fig. 4.2 (A) TEM of Fe_3O_4, (B) SEM of Al_2O_3, and (C) SEM of NiO nanoparticles [20].

calculates the amount of adsorbed gas corresponding to a monomolecular layer on the surface. Surface areas are calculated, using the BET equation, as given below [6, 21, 22]:

$$\frac{1}{V_a\left(\frac{P_a}{P}-1\right)}=\frac{C-1}{V_mC}\left(\frac{P}{P_0}\right)+\frac{1}{V_mC} \quad (4.6)$$

where V_a represents the volume of gas adsorbed at standard temperature and pressure (STP), V_m indicates the volume of gas adsorbed at STP to produce a monolayer on the sample surface, P_0 denotes the saturated pressure of adsorbate gas, P introduces the partial vapor pressure of the adsorbate gas in equilibrium with the surface at 77.4 K (boiling point temperature of liquid nitrogen), and C is a dimensionless constant that is related to the adsorption enthalpy of the gas and nanoparticle.

4.1.7.4 Nanoparticle size

Nanoparticle size becomes important in case of using the nanoparticles in porous media. Their size should be relatively smaller than pore throat to avoid plugging. The average crystallite size (d) of the nanoparticles is determined by applying the Scherrer equation to the main diffraction peaks of the XRD report [19, 21]. The Scherrer equation is given as follows:

$$d=\frac{K\lambda}{\beta\cos\theta} \quad (4.7)$$

where K is a shape factor, λ shows the wavelength of X-ray radiation, β denotes the line broadening at half of the maximum intensity of the diffracted peak, and θ represents the angle of diffraction [19]. In some cases, field-emission scanning electron microscopy is used to obtain more accurate values for particle sizes and the morphology of nanoparticles [6, 22–24]. In addition, dynamic light scattering (DLS) is conventionally used to measure the nanoparticle's size distribution [24].

4.1.7.5 Surface acidity of nanoparticles

The surface acidity of nanoparticles can help them to adsorb asphaltene more effectively if they are designed based on the asphaltene type. Ammonia temperature-programmed desorption (NH_3-TPD) is a useful technique to evaluate the relative amounts of acid sites on a nanoparticle surface due to the existent ammonia, which is a strong base (pKb $\simeq$ 5). In this test, a specific amount of the sample is introduced in a cylindrical glass TPD microreactor, which is degassed at 150°C overnight under a helium flow at atmospheric pressure. The sample is cooled to 100°C and NH_3 (10% in He) is introduced through the sample fixed bed for 1 h. Then, physisorbed NH_3 is removed by the helium flow over the solid in 1 h at 100°C. After that, the sample is heated to 900°C at a linear rate of 10 K/min for desorption of the chemisorbed NH_3. The difference of desorbed NH_3 versus temperature is calculated by the peak area using a calibration curve. Therefore, the amount of desorbed NH_3 is a measure of the nanoparticle acidity [22, 24].

4.1.8 Temperature tolerability

Temperature is a parameter that can affect the efficiency and also the stability of asphaltene inhibitors. Inhibitors like polymers may degrade at high temperatures. Thermal gravimetric analysis (TGA) is used as a degradation test. In this test, the temperature at which the polymer undergoes a major weight loss is considered as degradation temperature. The main goal of this test is to check whether the inhibitor can tolerate the reservoir conditions [25].

4.2 Thermodynamic behaviors of chemical inhibitors

Inhibitors are tested under different thermodynamic conditions. Often, the inhibitors are first tested at the standard conditions for simplicity. However, regarding the crude oil asphaltene deposition envelop, the asphaltene problem occurs under harsher operating conditions, including that in near wellbore, at bottom hole, and along the well; therefore, the asphaltene inhibitor efficiency should be verified at the actual thermodynamic conditions. Based on the experiments, an increase in the temperature usually improves the inhibitory efficiency of surfactants [26, 27], polymers [28, 29], and ionic liquids [30] (if they do not degrade), but reduces the inhibitory effect of zwitterionic liquids [31] and nanoparticles [23, 32–34]. For instance, Permsukarome et al. [26] investigated dodecylbenzenesulfonic acid (DBSA) in synthetic oil in the temperature range 6–58°C. The experimental results demonstrated that increasing temperatures improves the DBSA efficiency (Fig. 4.3). Moreover, Ogunlaja et al. [30] studied ionic liquids in the temperature range 25–90°C and showed that the ionic liquid inhibitory effect increases consistently upon an increase in temperature. They found 1-butyl-3-methylimidazolium with either chloride or nitrate to be effective at elevated temperatures; however, 1-methyl-1*H*-imidazol-3-ium-2-carboxybenzoate decomposed at higher temperatures.

The inhibitors are studied for crude oils under different pressures and temperatures, and are proven effective even at elevated pressures at lab and field scales [8, 35–39]. In some cases, it is mentioned that the effective concentration of inhibitors at elevated pressures and temperatures is less than that under standard conditions [37, 40]. Nevertheless, Gharbi et al. [41] claimed that the trend of inhibitor efficiency with pressure might fluctuate under different conditions. For instance, resins have less inhibitory effectiveness under ambient condition and better efficiency in pipeline or reservoir conditions. This observation is in line with the general trend that the

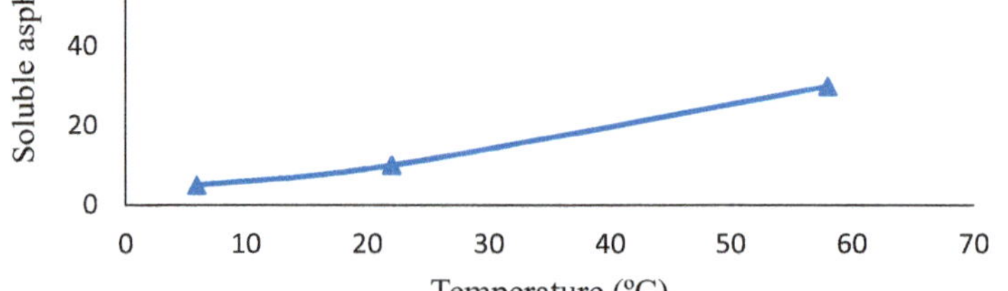

Fig. 4.3 Temperature efficiency in the presence/absence of DBSA at 20% dodecanol [26].

inhibitor efficiency improves when an increase in pressure and temperature occurs and the malfunction of the resins under ambient condition is justified by their self-aggregation [41].

The nanoparticle inhibitory behavior is different from other chemical inhibitors regarding the temperature impact. Due to the exothermic nature of the asphaltene-nanoparticle adsorption, its inhibitory efficiency decreases when temperature raises for metal nanoparticles [34] and metal oxide nanoparticles [23, 32]. Increasing temperature also lowers the inhibitors' affinity of other types of chemical inhibitors to adsorb onto the rock [33].

4.3 Stability of inhibitors

4.3.1 Effect of temperature

The stability of the inhibitors should be evaluated at operating temperatures. For example, 1-methyl-1*H*-imidazol-3-ium-2-carboxybenzoate loses its inhibitory effectiveness at 50°C [30] while some polymeric inhibitors can still function up to 250°C [25]. Moreover, some surfactants are only soluble in oil at elevated temperatures. For instance, heptyloxyphenol and dodecyl resorcinol are solubilized in alkanes above 50°C and 100°C, respectively [1, 42].

4.3.2 Effect of inhibitor concentration

The optimum inhibitor concentration can be defined in two different ways. Some inhibitors lose their efficiency by increasing the inhibitor concentration beyond the optimum concentration [43, 44]. Fig. 4.4 shows that increasing the DBSA concentration reduces its inhibitory efficiency. The second perspective of the inhibitor optimality is in terms of cost effectiveness; in this case, the inhibitor effectiveness still increases with concentration, but at a much lower rate [27, 45] or nearly zero [46], that it does not justify the additional cost. Sometimes, there is a minimum threshold concentration for the inhibitor to be effective [47]. At high concentrations, the inhibitors may cocrystallize, which may even intensify the asphaltene deposition problem. This issue usually occurs for polymers [48] and ionic inhibitors in nonionic media [10, 42, 47].

Controversy exists in the literature regarding the behavior and stability of different inhibitors, especially surfactants. For example, Lian et al. [14] observed that the inhibitory effect of nonylphenol (as a nonionic surfactant) increases with increasing concentration, whereas Permsukarome et al. [26]

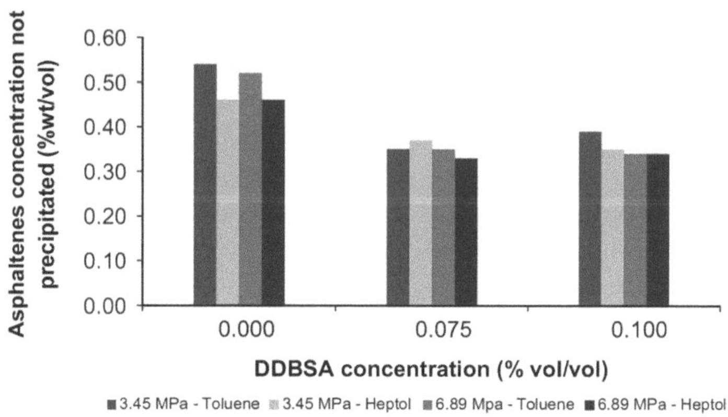

Fig. 4.4 DBSA lost its inhibitory efficiency when its concentration increased [43].

showed that its inhibitory effect decreases beyond an optimum concentration. In this case, the inconsistency lies in the differences among the experimental concentration ranges. In the study by Lian et al. [14], nonylphenol concentration range of 0%–1% was studied, while Permsukarome et al. [26] considered a wider range of 0%–24%; the optimum concentration was obtained at 7% which was more than the maximum value of 1% studied by Lian et al. [14]. As another example, Leon et al. [49] and Permsukarome et al. [26] have investigated DBSA as an asphaltene inhibitor in the range of 0%–18%; Permsukarome et al. [26] referred to 4% as the optimum concentration, while Leon et al. [49] found that DBSA even causes a higher level of efficiency above 4% concentration, which might be related to the media and asphaltene characteristics.

Organic inhibitors will not lose their efficiency with increasing concentration if they are effective and compatible with the problematic fluids. Organic inhibitors behave as natural inhibitors of asphaltene deposition and change the asphaltene deposition envelop of the target fluid. Although increasing the concentration of organic inhibitor does not cause more deposition, the rate of the improvement decreases after a specific concentration [5, 50].

4.3.3 Effect of intruder ions

The inhibitor's stability is different based on the inhibitor type and problematic hydrocarbon characteristics. For instance, DBSA, as an ionic inhibitor, can be effective in the presence of positively charged asphaltenes and even

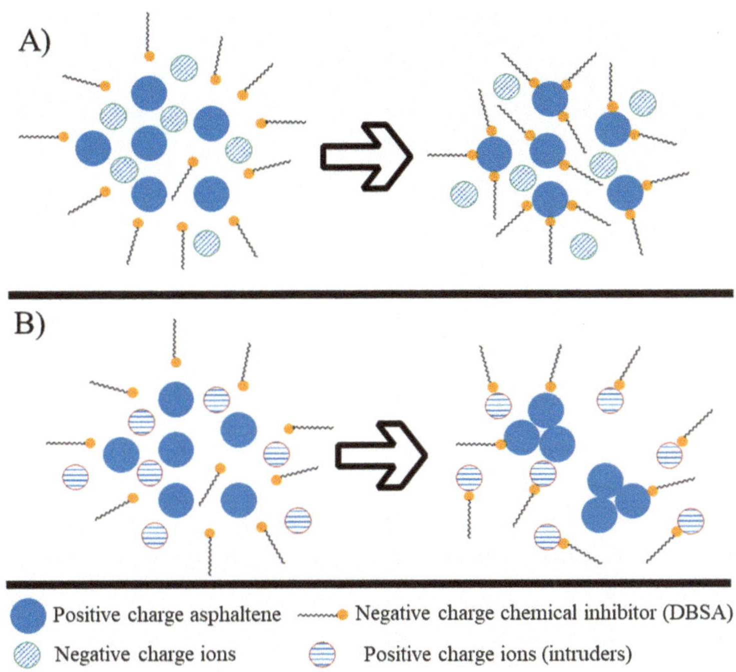

Fig. 4.5 (A) Inhibitor effect in the presence of noneffective ions and (B) inhibitor effect in the presence of intruder ions.

when noneffective ions exist (Fig. 4.5A). However, in the presence of intruder ions, if the concentration of DBSA is low, the DBSA interacts with the intruder ions and produces a salt instead of keeping the asphaltenes dispersed (Fig. 4.5B). Moreover, in acidic media or negatively charged asphaltene, DBSA does not show any inhibitory effect and, to some extent, DBSA can aggregate and make the deposition problem more severe [37, 51, 52]. Therefore, the asphaltene and oil chemistry and their compatibility with the selected inhibitor control the inhibitors' stability.

4.3.4 Effect of water and salinity

The presence of water (and brine) causes more deposition with ionic inhibitors such as DBSA, while water does not affect the efficiency of nonionic inhibitors. Usually, brine and water disrupt the ionic inhibitor's stability such that the effect of brine is less than that of water in the case of DBSA [53]. In addition, water can reduce the asphaltene adsorption capacity with nanoparticles such as alumina (Fig. 4.6) [32].

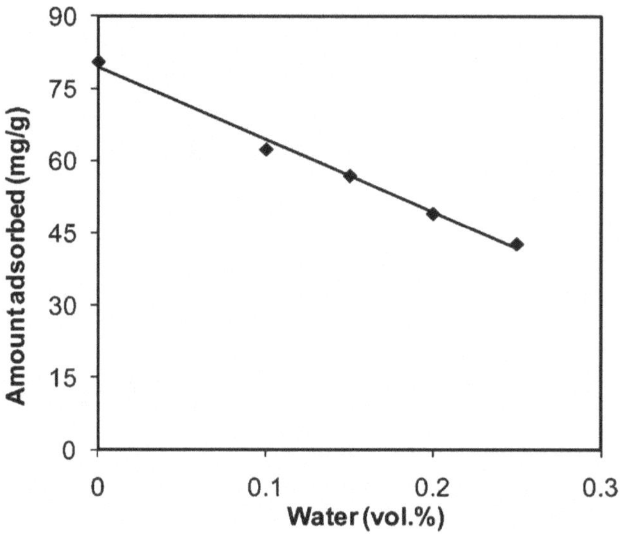

Fig. 4.6 Effect of water content on asphaltene adsorption on nanoparticles (γ-Al_2O_3) [32].

4.4 Final remarks

In this chapter, we presented the important parameters that affect the inhibitor efficiency. We also provided a brief introduction of the analytical methods to measure these parameters for different inhibitors. Moreover, the inhibitors' challenges with using synthetic, dead, and live crude oils were discussed under different thermodynamic conditions. For instance, nanoparticles' efficiency decreases by increasing temperature, while the efficiency of other inhibitor types may increase with temperature. Finally, the stability of inhibitors was discussed in the case of changing temperature, inhibitor concentration, and intruder ions and fluids. For instance, different types of inhibitors showed sensitivity to the presence of water and brine and/or their performance considerably changes with temperature.

References

[1] C.-L. Chang, H.S. Fogler, Stabilization of asphaltenes in aliphatic solvents using alkylbenzene-derived amphiphiles. 1. Effect of the chemical structure of amphiphiles on asphaltene stabilization, Langmuir 10 (6) (1994) 1749–1757.
[2] A.F. Lima, C.R. Mansur, E.F. Lucas, G. González, Polycardanol or sulfonated polystyrene as flocculants for asphaltene dispersions, Energy Fuel 24 (4) (2009) 2369–2375.
[3] L.F.B. Moreira, E.F. Lucas, G. González, Stabilization of asphaltenes by phenolic compounds extracted from cashew-nut shell liquid, J. Appl. Polym. Sci. 73 (1) (1999) 29–34.

[4] A.M. Atta, A.O. Ezzat, M.M. Abdullah, A.I. Hashem, Effect of different families of hydrophobic anions of imadazolium ionic liquids on asphaltene dispersants in heavy crude oil, Energy Fuel 31 (8) (2017) 8045–8053.
[5] O. Leon, E. Contreras, E. Rogel, G. Dambakli, J. Espidel, S. Acevedo, The influence of the adsorption of amphiphiles and resins in controlling asphaltene flocculation, Energy Fuel 15 (5) (2001) 1028–1032.
[6] M. Mohammadi, M. Dadvar, B. Dabir, TiO_2/SiO_2 nanofluids as novel inhibitors for the stability of asphaltene particles in crude oil: mechanistic understanding, screening, modeling, and optimization, J. Mol. Liq. 238 (2017) 326–340.
[7] M. Barcenas, P. Orea, Molar-mass distributions of asphaltenes in the presence of inhibitors: experimental and computer calculations, Energy Fuel 25 (5) (2011) 2100–2108.
[8] E.F. Ghloum, M. Al-Qahtani, A. Al-Rashid, Effect of inhibitors on asphaltene precipitation for Marrat Kuwaiti reservoirs, J. Petrol. Sci. Eng. 70 (1) (2010) 99–106.
[9] N. Da Silva, C. Batistella, R. Maciel Filho, M. Wolf Maciel, Determination of castor oil molecular weight by vapour pressure osmometry technique, Chem. Eng. Trans. 24 (2011) 601–606.
[10] M. Barcenas, P. Orea, E. Buenrostro-González, L.S. Zamudio-Rivera, Y. Duda, Study of medium effect on asphaltene agglomeration inhibitor efficiency, Energy Fuel 22 (3) (2008) 1917–1922.
[11] K. Akbarzadeh, et al., Asphaltenes—problematic but rich in potential, Oilfield Rev. 19 (2) (2007) 22–43.
[12] M. Mousavi, T. Abdollahi, F. Pahlavan, E.H. Fini, The influence of asphaltene-resin molecular interactions on the colloidal stability of crude oil, Fuel 183 (2016) 262–271.
[13] V.L. Weinberg, T.F. Yen, 'Solubility parameters' in coal and coal liquefaction products, Fuel 59 (5) (1980) 287–289.
[14] H. Lian, J.-R. Lin, T.F. Yen, Peptization studies of asphaltene and solubility parameter spectra, Fuel 73 (3) (1994) 423–428.
[15] R.D. Shannon, Revised effective ionic radii and systematic studies of interatomic distances in halides and chalcogenides, Acta Crystallogr. Sect. A: Cryst. Phys., Diffr., Theor. Gen. Crystallogr. 32 (5) (1976) 751–767.
[16] Y. Iwadate, K. Kawamura, K. Igarashi, J. Mochinaga, Effective ionic radii of nitrite and thiocyanate estimated in terms of the Boettcher equation and the Lorentz-Lorenz equation, J. Phys. Chem. 86 (26) (1982) 5205–5208.
[17] S. Zhang, N. Sun, X. He, X. Lu, X. Zhang, Physical properties of ionic liquids: database and evaluation, J. Phys. Chem. Ref. Data Monogr. 35 (4) (2006) 1475–1517.
[18] D. Subramanian, K. Wu, A. Firoozabadi, Ionic liquids as viscosity modifiers for heavy and extra-heavy crude oils, Fuel 143 (2015) 519–526.
[19] M. Mohammadi, M. Akbari, Z. Fakhroueian, A. Bahramian, R. Azin, S. Arya, Inhibition of asphaltene precipitation by TiO_2, SiO_2, and ZrO_2 nanofluids, Energy Fuel 25 (7) (2011) 3150–3156.
[20] F. Shojaati, M. Riazi, S.H. Mousavi, Z. Derikvand, Experimental investigation of the inhibitory behavior of metal oxides nanoparticles on asphaltene precipitation, Colloids Surf. A Physicochem. Eng. Asp. 531 (2017) 99–110.
[21] C.A. Franco, N.N. Nassar, M.A. Ruiz, P. Pereira-Almao, F.B. Cortés, Nanoparticles for inhibition of asphaltenes damage: adsorption study and displacement test on porous media, Energy Fuel 27 (6) (2013) 2899–2907.
[22] N.N. Nassar, A. Hassan, P. Pereira-Almao, Effect of the particle size on asphaltene adsorption and catalytic oxidation onto alumina particles, Energy Fuel 25 (9) (2011) 3961–3965.
[23] N. Hosseinpour, A.A. Khodadadi, A. Bahramian, Y. Mortazavi, Asphaltene adsorption onto acidic/basic metal oxide nanoparticles toward in situ upgrading of reservoir oils by nanotechnology, Langmuir 29 (46) (2013) 14135–14146.

[24] S. Betancur, J.C. Carmona, N.N. Nassar, C.A. Franco, F.B. Cortés, Role of particle size and surface acidity of silica gel nanoparticles in inhibition of formation damage by asphaltene in oil reservoirs, Ind. Eng. Chem. Res. 55 (21) (2016) 6122–6132.
[25] M. Wang, J. Kaufman, X. Chen, C. Sungail, Development and evaluation of non-ionic polymeric surfactants as asphaltene inhibitors, in: SPE International Symposium on Oilfield Chemistry, Society of Petroleum Engineers, 2015.
[26] P. Permsukarome, C. Chang, H.S. Fogler, Kinetic study of asphaltene dissolution in amphiphile/alkane solutions, Ind. Eng. Chem. Res. 36 (9) (1997) 3960–3967.
[27] M.A. Karambeigi, M. Nikazar, R. Kharrat, Experimental evaluation of asphaltene inhibitors selection for standard and reservoir conditions, J. Petrol. Sci. Eng. 137 (2016) 74–86.
[28] C. Ovalles, E. Rogel, H. Morazan, K. Chen, M.E. Moir, The use of nonylphenol formaldehyde resins for preventing asphaltene precipitation in vacuum residues and hydroprocessed petroleum samples, Pet. Sci. Technol. 34 (4) (2016) 379–385.
[29] C. Ovalles, E. Rogel, H. Morazan, M.E. Moir, Synthesis, characterization, and mechanism of asphaltene inhibition of phosphopropoxylated asphaltenes, Fuel 180 (2016) 20–26.
[30] A.S. Ogunlaja, E. Hosten, Z.R. Tshentu, Dispersion of asphaltenes in petroleum with ionic liquids: evaluation of molecular interactions in the binary mixture, Ind. Eng. Chem. Res. 53 (48) (2014) 18390–18401.
[31] L.A. Alcázar-Vara, L.S. Zamudio-Rivera, E. Buenrostro-González, R.L. Hernández-Altamirano, V.Y. Mena-Cervantes, J.F. Ramírez-Pérez, Multifunctional properties of Zwitterionic liquids. Application in enhanced oil recovery and Asphaltene aggregation phenomena, Ind. Eng. Chem. Res. 54 (11) (2015) 2868–2878.
[32] N.N. Nassar, Asphaltene adsorption onto alumina nanoparticles: kinetics and thermodynamic studies, Energy Fuel 24 (8) (2010) 4116–4122.
[33] E. Joonaki, R. Burgass, B. Tohidi, Experimental and modelling study on application of new class of asphaltene inhibitors for enhanced oil recovery purposes: adsorption and wettability alteration, in: Abu Dhabi International Petroleum Exhibition & Conference, Society of Petroleum Engineers, 2016.
[34] H. Alboudwarej, D. Pole, W.Y. Svrcek, H.W. Yarranton, Adsorption of asphaltenes on metals, Ind. Eng. Chem. Res. 44 (15) (2005) 5585–5592.
[35] Y.-F. Hu, T.-M. Guo, Effect of the structures of ionic liquids and alkylbenzene-derived amphiphiles on the inhibition of asphaltene precipitation from CO_2-injected reservoir oils, Langmuir 21 (18) (2005) 8168–8174.
[36] A.A. Melendez-Alvarez, et al., On the evaluation of the performance of asphaltene dispersants, Fuel 179 (2016) 210–220.
[37] E. Zahedi, M. Vafaie-Sefti, M. Shadman, H. Naderi, M. Amiri, A. Noorbakhsh, Experimental investigation of dodecylbenzene sulfonic acid and toluene dispersants on asphaltene precipitation of dead and live oil, Pet. Sci. Technol. 35 (7) (2017) 653–660.
[38] L.M. Cenegy, Survey of successful world-wide asphaltene inhibitor treatments in oil production fields, in: SPE Annual Technical Conference and Exhibition, Society of Petroleum Engineers, 2001.
[39] Y. Villard, F. Fajardo, A. Milne, Enhanced oil recovery using innovative asphaltene inhibitors in East Venezuela, in: SPE International Conference and Exhibition on Formation Damage Control, Society of Petroleum Engineers, 2016.
[40] M. Bouts, R. Wiersma, H. Muijs, A. Samuel, An evaluation of new aspaltene inhibitors; laboratory study and field testing, J. Petrol. Tech. 47 (09) (1995) 782–787.
[41] K. Gharbi, K. Benyounes, C. Benamara, Inhibition of asphaltene flocculation in light crude oil: effect of pressure, temperature and inhibitor concentration, Pet. Sci. Technol. 36 (18) (2018) 1471–1477.

[42] C.-L. Chang, H. Fogler, Asphaltene stabilization in alkyl solvents using oil-soluble amphiphiles, in: SPE International Symposium on Oilfield Chemistry, Society of Petroleum Engineers, 1993.
[43] T.S. Loureiro, L.C.M. Palermo, L.S. Spinelli, Influence of precipitation conditions (n-heptane or carbon dioxide gas) on the performance of asphaltene stabilizers, J. Petrol. Sci. Eng. 127 (2015) 109–114.
[44] S. Kashefi, A. Shahrabadi, S. Jahangiri, M.N. Lotfollahi, H. Bagherzadeh, Investigation of the performance of several chemical additives on inhibition of asphaltene precipitation, Energy Source A: Recov Util Environ Effects 38 (24) (2016) 3647–3652.
[45] T. Lu, Z. Li, W. Fan, X. Zhang, Q. Lv, Nanoparticles for inhibition of asphaltenes deposition during CO_2 flooding, Ind. Eng. Chem. Res. 55 (23) (2016) 6723–6733.
[46] Y. Kazemzadeh, S.E. Eshraghi, K. Kazemi, S. Sourani, M. Mehrabi, Y. Ahmadi, Behavior of asphaltene adsorption onto the metal oxide nanoparticle surface and its effect on heavy oil recovery, Ind. Eng. Chem. Res. 54 (1) (2015) 233–239.
[47] A. Aquino-Olivos, E. Buenrostro-Gonzalez, S.I. Andersen, C. Lira-Galeana, Investigations of inhibition of asphaltene precipitation at high pressure using bottomhole samples, Energy Fuel 15 (1) (2001) 236–240.
[48] M. Alhreez, D. Wen, L. Ali, A novel inhibitor for controlling Iraqi asphaltene problems, in: Environmental Impacts of the Oil and Gas Industries: Kurdistan Region of Iraq as a Case Study (EIOGI), 2017 International Conference on, IEEE, 2017, pp. 37–41.
[49] O. León, E. Rogel, A. Urbina, A. Andújar, A. Lucas, Study of the adsorption of alkyl benzene-derived amphiphiles on asphaltene particles, Langmuir 15 (22) (1999) 7653–7657.
[50] I. Auflem, T. Havre, J. Sjöblom, Near-IR study on the dispersive effects of amphiphiles and naphthenic acids on asphaltenes in model heptane-toluene mixtures, Colloid Polym. Sci. 280 (8) (2002) 695–700.
[51] J. Wang, C. Li, L. Zhang, G. Que, Z. Li, The properties of asphaltenes and their interaction with amphiphiles, Energy Fuel 23 (7) (2009) 3625–3631.
[52] L. Goual, A. Firoozabadi, Effect of resins and DBSA on asphaltene precipitation from petroleum fluids, AICHE J. 50 (2) (2004) 470–479.
[53] D. Subramanian, A. Firoozabadi, Effect of surfactants and water on inhibition of asphaltene precipitation and deposition, in: Abu Dhabi International Petroleum Exhibition and Conference, Society of Petroleum Engineers, 2015.

CHAPTER FIVE

Experimental investigation of asphaltene deposition control by chemical inhibitors

5.1 Experimental methods

Before any field implementation, the inhibitors should be evaluated at the reservoir conditions in a laboratory setup to screen potential asphaltene inhibitors. However, simulating operational field conditions in the lab is normally difficult or impractical to achieve. Therefore, different types of tests are performed to evaluate the asphaltene-inhibitor interactions under both static and dynamic conditions. Even in some cases, diagnostic workflows are introduced to find the best inhibitor by coupling several testing methods [1, 2]. The conventional techniques are categorized in Fig. 5.1.

5.1.1 Static methods

Static methods of assessing the asphaltene precipitation inhibition are the methods that analyze a stagnant fluid sample. In most cases, the inhibitor effectiveness is initially established through primitive methods such as direct observation where the sample behavior is analyzed in a test vial. Because these tests are usually performed at atmospheric condition with simple instruments, they are capable to assess multiple inhibitors (with different concentrations). Conducting static tests, it is possible to reduce the number of experimental variables for more vigorous tests such as flow loop in tubes or porous media. The static tests do not account for the flow hydrodynamics and fluid-shearing effects. There are different apparatuses and procedures designed for static tests, including spectroscopic, visual, and thermal methods that can show inhibitor effectiveness.

5.1.1.1 Spectrophotometric

The spectrophotometric techniques are based on the emission of light at a specific wavelength and measuring the light absorbance through the sample.

Asphaltene Deposition Control by Chemical Inhibitors
https://doi.org/10.1016/B978-0-323-90510-7.00007-0

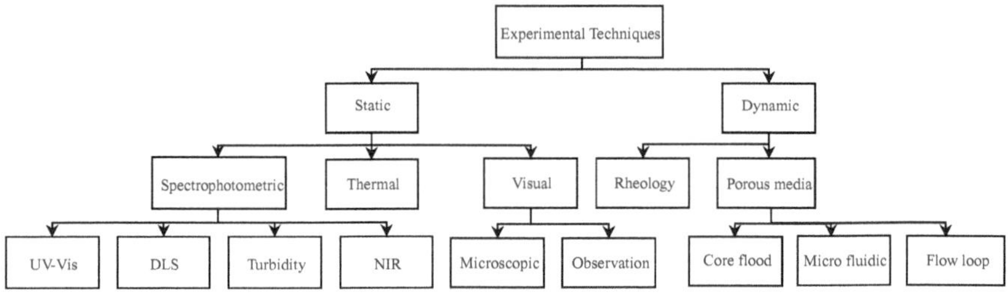

Fig. 5.1 Classification of inhibitor evaluation techniques.

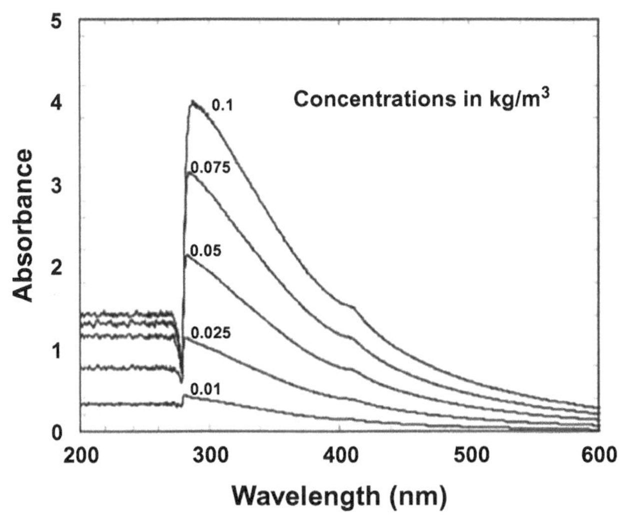

Fig. 5.2 UV-vis absorbance spectra of low concentration solutions of Athabasca *n*C7 asphaltene in toluene at 22°C [3].

Depending on the sample, an efficient wavelength is identified by testing a range of wavelengths (Fig. 5.2) [3]. Based on the emitted light wavelength, the spectrophotometric test can be categorized under ultraviolet-visible (UV-vis) [3–17], dynamic light scattering (DLS) [6, 14, 18–22], turbidity [8, 9, 23–28], and Near-Infrared (NIR) [29–33], which are discussed in the following sections.

Ultraviolet-visible (UV-vis)

UV-vis spectroscopy measures the light absorbance to evaluate the asphaltene aggregation, dispersion, and inhibition after the addition of precipitant (normally *n*C7) to the solution [10]. The UV-vis experiment is based on the insolubility of asphaltene in a precipitant, and the consequent stability of the asphaltene in the presence of inhibitors [10]. In UV-spectroscopy, asphaltene dissolved in toluene is considered as a reference sample. The inhibitors are then added to the reference mixture, at a range of the inhibitor concentrations, with a fixed amount of the precipitant. The samples are usually mixed using a vortex system, following centrifugation, and settling the sample to stabilize. Then, a specific volume of the supernatant sample is diluted with toluene. The diluted supernatant's behavior is observed in a quartz cell with an optical path range of the 2–100 mm, using the light wavelength of 190–410 nm [3–17]. The inhibitor efficiency is obtained by calculating the ratio of the test sample absorbance (A_{test}) to the reference sample (A_{ref}), as shown by Eq. (5.1) [10]:

$$\%\,\text{Efficiency} = \frac{A_{test}}{A_{ref}} \times 100 \tag{5.1}$$

The effect of inhibitor dilution is compensated by subtracting its effect on absorbance from samples [10]. The schematic representation of a UV/UV-vis system is shown in Fig. 5.3 to make the process more sensible [34].

Dynamic light scattering (DLS)

The dynamic light-scattering (DLS) technique uses a linear relationship between the concentration of asphaltene in a solvent and the light absorbance [7, 35]. The DLS measurement is usually conducted using a Zetasizer (Malvern Panalytical Ltd., UK) and a ZetaPALS (Brookhaven Instruments, USA) as a NanoBrook Series Particle Analyzer [19]. In the DLS technique, the dynamic growth and size of the asphaltene particles as well as their surface charges are measured. In this test, the effects of colloid stability,

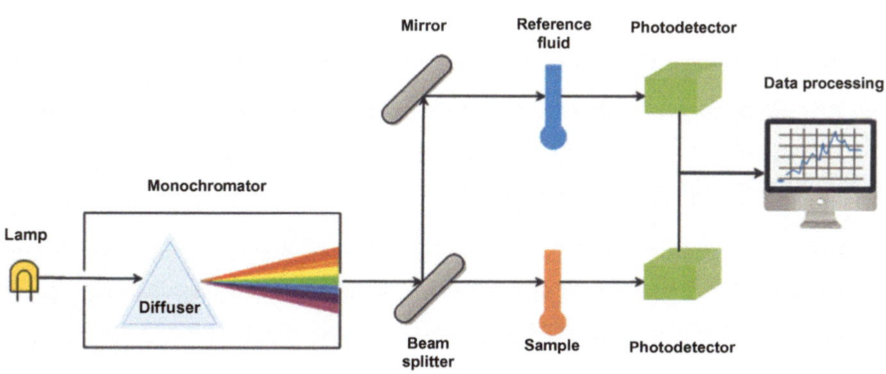

Fig. 5.3 A schematic representation of a UV or UV-vis system [34].

aggregation rate, and interparticle interactions are considered [20]. Using the DLS technique to investigate the colloidal aggregates results in a more comprehensive understanding of diffusion and reaction-limited or diffusion aggregation processes [36]. The DLS method is capable of measuring the aggregation of asphaltene in dark samples of heavy oils (and, up to 7 wt% asphaltene mixed with toluene) [19, 37, 38]. Increasing the toluene content from 75 to 95 vol% causes the asphaltene aggregates to grow in 70–7000 min from submicrometer to 10 μm [39]. A specific volume of nC_7 is mixed with various amounts of oil in centrifuge tubes. After sonicating the sample, a specific volume of the mixture is immediately transferred to a cuvette for size and zeta potential measurements [19]. Given the asphaltene content (f) and asphaltene density ($\tilde{n}_a$), with a 3-cm sample height, the measurement is applied to the top 1 cm of the sample, and none of the samples used for DLS exceeds a volume fraction of asphaltenes $\varphi \sim \frac{f}{\rho_a X} \sim 0.0005$, where φ refers to the volume fraction, and X is the ratio of heptane (in mL) per oil (in g) [19]. The duration of the DLS tests varies according to the sample's composition and scattered light intensity, I, which is the number of scatters in the light paths, used in data analysis [40].

Turbidity

Turbidity is another type of spectroscopic test that measures the clarity of a fluid as a measure of asphaltene deposition. The fluid expansion refers to the dilution in volume and number of molecules, which results in decreasing the light absorbance per unit volume of the sample, thereby increasing the light transmission. However, an increase in the fluid turbidity leads to a decrease in light transmission as a result of the particle aggregation [24]. The turbidity system (such as a Turbiscan) relies on not only the particle concentration but also the particle size and distribution [24]. In this method, photons are emitted into the sample and are scattered by the suspended particles, and asphaltene aggregates are detected through the transmission and backscatter detectors [24]. The Turbiscan device measures changes in transmitted light through the oil sample (that includes asphaltene, solvent, precipitant, and dispersant), as a result of asphaltene aggregation [26]. Observing a change in the transmitted light after adding the inhibitor reveals a high inhibitor effectivity [26]. This technique quantitatively evaluates the asphaltene stability in crude oil (or asphaltene solvent) using the separability number (%). Separability numbers greater than 10 refer to unstable samples; separability numbers less than 5 refer to a stable sample; and separability numbers between 5 and 10 refer to medium stability [26].

Near-infrared (NIR)

The Near-Infrared (NIR) spectroscopy method is a reliable, accurate, and fast technique to determine the initial screening of a large number of asphaltene inhibitors [31, 33]. The NIR spectral range is defined as a relative amount of light absorption and scattering by the asphaltene particles in the crude oils [31]. The NIR spectroscopy investigates the dissociation of the asphaltene aggregation at specific wavelengths [33]. This technique is also applied to determine the efficiency of various additives to dissociate the asphaltene aggregates. The NIR measurements are performed with a spectrometer equipped with a fiber optic sensor, which is used for transparency measurements [33]. The wavelength range and the total number of scans per spectrum required in the NIR technique are in the range of 750–2500 nm [29–32] and 30–40 [33], respectively. Usually, the process is conducted at room temperature, and a total sample path length of 1 mm is selected. The light scattering in crude oil samples depends on the size and morphology of the heavy particles, such as asphaltenes, wax, and hydrates [33]. Measuring the relative transmitted light through a sample is a common method to obtain a spectroscopic spectrum. In this method, the transmitted light is converted into the optical density (OD) through a nonlinear relation [Eq. (5.2)] [31, 33].

$$\mathrm{OD} = \log\left(\frac{I_0}{I}\right) = 0.434N\sigma_{tot} \tag{5.2}$$

OD is the amount of light lost or absorbed by the crude oil sample, and I_0 and I introduce the intensity of the adsorbed and transmitted light, respectively. σ_{tot} is the total cross section containing N particles. Auflem et al. [33] used NIR scattering measurements at 1600 nm for 0.125 wt% asphaltene solution in a 70/30 *n*-heptane/toluene mixture with 1.25 wt% of various inhibitors, and measured asphaltene precipitation.

5.1.1.2 Thermal techniques

Theoretically, the thermal resistance of an oil mixture to transfer heat increases when asphaltene precipitates [41]. Fig. 5.4 shows a schematic representation of the asphaltene precipitation test cell that was used by Clark and Pruden at 400°C and 2000 psi [41, 42]. In this device, heat is provided with the heating tapes around the vessel walls to facilitate conduction heat transfer in the deposited asphaltene layer at the bottom of the vessel. The cell is also isolated from the top and bottom using gaskets, and both top plate and vessel wall are held at a constant temperature [42]. In the heat transfer experiments, an oil sample with/without the inhibitor is weighed and placed into

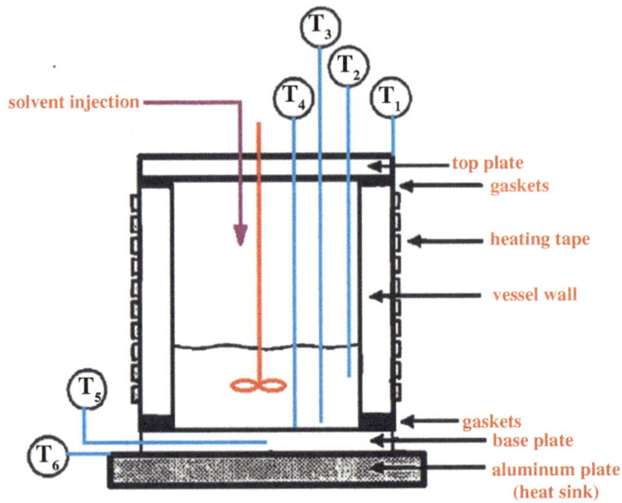

Fig. 5.4 Schematic representation of heat transfer cell for determining the onset of asphaltene precipitation [41].

the cell [42]. The temperature of the vessel remains constant. The sample is heated and in each cycle, the stirrer is "turned on" for 5 min, then "turned off" to provide a steady-state profile of temperature [42]. In the last 10 cycles, a constant volume of nC_7 or toluene is injected into the vessel and in the last 10 min of each cycle, the average temperature is recorded to obtain the average steady-state temperature in each cycle [42]. In the case of adding precipitant to the oil sample, the viscosity of the mixture decreases, and the temperature of the bottom part of the cell increases when the ratio of precipitant to oil reaches the asphaltene onset point. The deposited layer at the bottom of the cell creates a new thermal resistance that results in temperature reduction at the bottom region of the vessel, which is used to quantify asphaltene deposition [42].

5.1.1.3 Visual methods

Visual techniques allow to observe and record the asphaltene aggregation or dispersion with naked eye, camera, and/or microscope. The asphaltene dispersant test, interfacial tension measurement, optical microscopy, and high-resolution transmission electron microscopy are among the popular techniques that are described in the following subsections.

Asphaltene dispersant test (ADT)

The asphaltene dispersant test (ADT) is economical and simple which is normally used to evaluate the inhibitor's activity, based on the insolubility of the asphaltene in a precipitant such as nC_7 [43]. According to the procedure proposed by Manek [44], the ADT method is conducted in two steps, as given below.

Preliminary step (A):

(1) 10 mL of hexane is added to each centrifuge tube.

(2) The crude oil is diluted with toluene at a 1:1 (vol) dilution ratio. Then, various amounts of oil (50, 100, 200, and 300 μL) are added to the tubes and allowed to stabilize for 2 h. The height of asphaltene deposition by the gravity force is recorded and crudes with 4%–10% deposition are selected for step (B).

Second step (B):

(1) Solution samples with 1 wt% dispersant are prepared.

(2) 10 mL of hexane is added to each centrifuged tube.

(3) The tubes are filled with desired volumes of dispersant(s) and mixed thoroughly.

(4) Based on the selected doses in step A, a proper volume of the crude oil is added to each tube, stirred for 60 s, and left for 2 h to stabilize. Again, the amount of precipitation by gravity force is recorded, and the effectiveness of inhibitor is determined by the following equation:

$$\%\,\text{Dispersant} = \left(1 - \frac{S_t}{S_b}\right) \times 100 \tag{5.3}$$

where S_t refers to the amount of precipitation in the treated sample, and S_b denotes the amount of precipitation in the bulk volume.

Interfacial tension (IFT) method

The IFT or vanishing interfacial tension (VIT) technique (see Fig. 5.5) measures the onset and intensity of asphaltene precipitation upon gas injection [45]. The disappearance of an interface between the two phases in terms of a measurable variable implies zero IFT between the two phases [46]. This technique is accurate and affordable and uses only small sample volumes in each test. The asphaltene precipitation onset is detected by measuring the IFT between the oil and the precipitant and analyzing the change in IFT slope in terms of the pressure of the starting point and intensity that

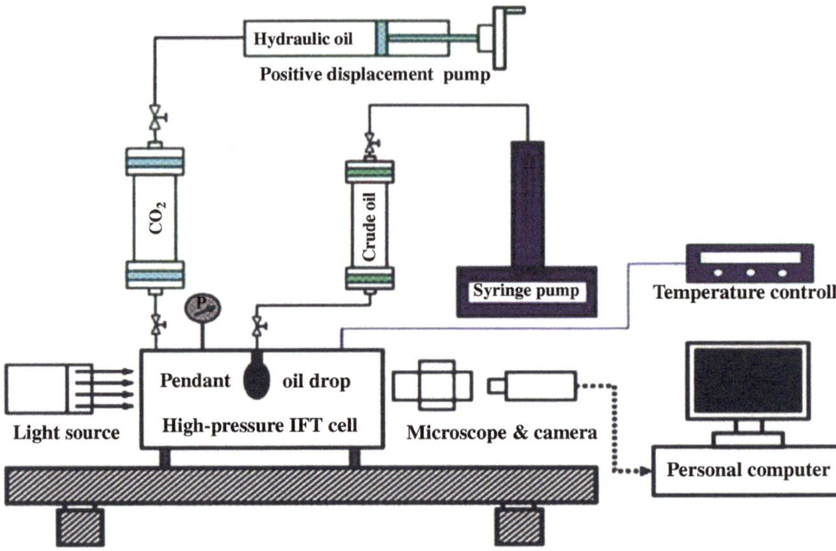

Fig. 5.5 Sketch of the experimental setup [45].

are two factors to validate the inhibitor efficiency [47–49]. Once the precipitation starts, the slope of IFT versus pressure changes sharply [45]. The following procedure is usually performed to measure the IFT at different pressures and temperatures:

- Clean the apparatus with solvents such as an alcohol and toluene before each test, purge with N_2 and rapidly evacuate, to avoid contaminations.
- When CO_2 is considered as a precipitant, the cell is saturated with pressurized CO_2 and is allowed to equilibrate at a given pressure and temperature.
- A droplet of the crude oil is injected from the oil tank into the view cell. The repeatability of these results should be checked by testing several droplets to reduce the error.
- Image processing software is used to capture the shape of the droplet continuously to calculate IFT.

During the analysis of the IFT measurements, the change in the IFT values with pressure gives information regarding the asphaltene precipitation onset. For instance, when using CO_2 as the precipitant, the crude oil-CO_2 IFT curves versus pressure depict different regions with different slopes. At low pressures, the IFT value reduction is related to the vaporization of light components from the crude oil (vaporizing) and the gas exsolving into the oil droplet. However, the slope change at higher pressures is due to the precipitation of heavy and large molecules of asphaltene at the interface of two phases. The pressure gradient in this region is caused by the perturbation in the intermolecular force balance, followed by a sudden increase in the IFT [45].

Coupon deposition test

The coupon deposition test is a method that measures the rate of asphaltene (in general scale) deposition visually and qualitatively. In this test, slices of stainless steel, which are scaled with different types and sizes of holes, rotate in the sample for a specific time, usually 4–7 days. Then, the slice is inspected visually to inspect the plugged and open holes. Therefore, the efficiency of inhibitors in the prevention of asphaltene deposition is measured by comparing the size and the type of the holes that are plugged after conducting a test in the sample with/without inhibitor [50].

Optical microscopy technique

The optical microscopy technique is widely used in asphaltene research studies to determine the asphaltene precipitation and deposition onset, in which the oil sample under the ambient conditions is titrated through increasing the concentrations of precipitant [51–54]. In this technique, an optical microscope system monitors the oil titration with a precipitant [55]. According to Wang and Buckley [56], the asphaltene aggregates grow continuously [55]. The asphaltene precipitation onset occurs at a threshold point where the aggregated particles grow to a size that is detectable by an optical microscope. In the literature, the typical threshold range for the asphaltene particle is 0.2–0.5 μm, depending on the microscope resolution; this is a reasonable range compared to the reservoir pore size [55]. The residence time and concentration of precipitant are the most influential parameters when using optical microscopy. This method is conventionally used in the investigation of asphaltene adsorption on nanoparticle inhibitors. Fig. 5.6 depicts pictures of asphaltene adsorption on different nanoparticles taken by scan electron microscopy (SEM). The nanoparticles that are used to adsorb asphaltene include NiO, Fe_2O_3, WO_3, MgO, $CaCO_3$, and ZrO_2 [57].

The optical microscopy technique is also employed to visualize the asphaltene precipitation/inhibition process from a live crude sample [55]. In this application, the crude sample is depressurized using a high-pressure microscope (HPM) device, which includes a high pressure/temperature cell as well as a light source and a camera. The images captured by the camera are

Fig. 5.6 FESEM micrographs of different nanoparticles before (left) and after (right) asphaltene adsorption: (A) NiO, (B) Fe_2O_3, (C) WO_3, (D) MgO, (E) $CaCO_3$, and (F) ZrO_2 [15]. The scale bars are 500 nm in every picture except in the left picture of panel E, which is 1 μm.

received by a data acquisition system and analyzed with the image processing tool [55]. The accuracy of optical microscopy depends on the opacity of the oil sample. This technique, in general, yields more reliable and accurate results for oil samples with lower gravity and low asphaltene content [55].

High-resolution transmission electron microscopy (HR-TEM)

In the high-resolution transmission electron microscopy (HR-TEM) technique, images are captured with molecular-scale details. Two or more beams of electron diffraction patterns are located in the back focal plane of the objective lens of a TEM connector [58]. The HR-TEM images verify the presence of aromatic ring system around ~1 nm diameter for crude oil asphaltene as well as 0.7 nm for coal asphaltene. The images can be directly interpreted to find the asphaltene-inhibitor interactions; however, in some cases, there is vagueness in the interpretation of images due to high thickness of material or image deficiencies. The HR-TEM imaging system uses very thin samples that partly transmit the electron beam. In poorly ordered structures, the thickness of the sample is the most probable source of error due to the difficulties of eliminating the superposition of the lattice fringe [59].

In this technique, asphaltene is dissolved in solutions with different ratios of nC_7 and toluene (heptol) to capture the size of the asphaltene aggregates with and without inhibitors, such as octylphenol [60]. The samples are allowed to stabilize for 1 day before analyzing. A thin grid (~15 nm) of silicon nitride (Si_3N_4) containing several windows of dimensions 0.1 (mm) × 0.1 (mm) is dipped horizontally into each sample for 30 s, allowing it to dry; images are then captured using a microscope (usually 200 kV accelerating voltage). An image processing software such as ImageJ needs to be used for analyzing the captured images [60]. Fig. 5.7 shows the results of HR-TEM micrograph at different ratios of heptol solutions.

5.1.2 Dynamic analysis

After evaluating the effectiveness of inhibitors under static conditions, the effectiveness tests should be conducted under dynamic conditions to account for the shear effects. The dynamic tests are categorized to the rheology tests such as viscosity measurements which represent crude behavior in pipelines, cores, and packed beds, and microfluidics tests which represent crude behavior in porous media. The onset point measurement through viscosity tests for samples with and without an inhibitor, shows inhibitor efficiency in dynamic conditions. The efficiency of the inhibitors in porous media (and pipelines) is determined by measuring the pressure drop, flow

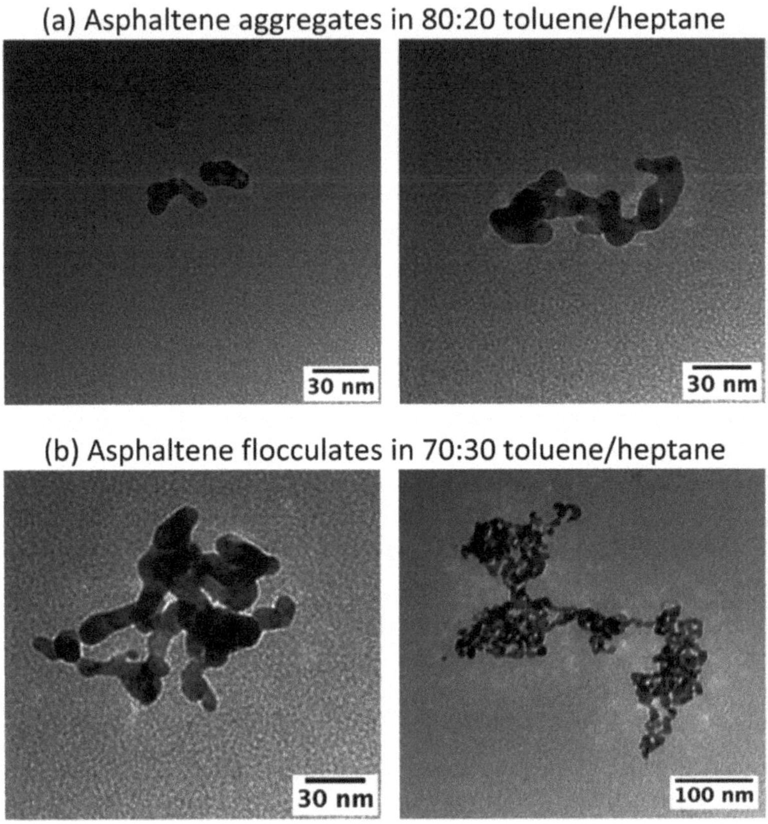

Fig. 5.7 HR-TEM micrographs of 5000 ppm asphaltenes in (A) 80:20 and (B) 70:30 heptol solutions imaged on Si_3N_4 grids [60].

rate, permeability, asphaltene content in the effluent, and asphaltene damage to the porous media. In microfluidics, the asphaltene precipitation is also visually detectable. The dynamic methods are described in the following section of this chapter.

5.1.2.1 Rheology techniques

The viscosity of a suspension system is controlled by the fluid-particle and particle-particle interactions, and the shape, concentration, and size distribution of the particles [55]. In colloidal systems, the viscosity is related to the asphaltene precipitation onset point [61–65]. This technique is designed based on considering the asphaltene mixtures as "colloidal suspensions" [55]. In this technique, asphaltenes are assumed to be stabilized by resins (on their surfaces); the aggregation of suspended asphaltene particles consequently increases the viscosity of the fluid [55]. The viscosity measurement is conducted during the precipitation test to detect the onset point [55]. Normally, the viscosity of the oil sample diluted with a precipitant is measured at a certain temperature. The onset point for asphaltene precipitation is determined by plotting viscosity versus the precipitant concentration with and without the inhibitor. Then, the results for samples with different concentrations of inhibitor(s) should be compared to evaluate inhibitor efficiency in various conditions [66]. Fig. 5.8 shows sample results of apparent viscosity measurements with and without the addition of DBSA (as asphaltene inhibitor) at various temperatures.

5.1.2.2 Porous media

Coreflood technique: Coreflood is a technique that evaluates the fluid/rock properties in a reservoir rock sample. The coreflood test assesses the asphaltene inhibitors under dynamic conditions in porous media; the operating conditions represent hydrocarbon reservoir conditions [4, 28, 68–72]. Major parts of a typical coreflood test include core holder, high-pressure positive displacement pumps, transfer vessels, controlled temperature environment, pressure transducers, flowmeter, and back pressure regulator [71]. To assess the effectiveness of asphaltene inhibitors within the coreflood setup, a known volume of inhibitor is accumulated into a transfer vessel and pressurized to the injection pressure. Based on the process design, the inhibitor is injected simultaneously or sequentially. After each experiment, both input fluid and output fluid of the core are analyzed through tests such as SARA, DLS, and/or SEM to investigate the behavior of asphaltene particles [71]. In addition, variables such as pressure drop, permeability, total fluid injection, and flow rate are used to assess the inhibitor efficiency on asphaltene deposition [17, 73, 74].

Microfluidics: In the microfluidic method, transparent microfluidic chips with different geometries and pore structures are used. The microfluidics can be used to study the mechanisms of asphaltene precipitation and deposition with and without an inhibitor [75]. Microfluidic tests are conducted either at atmospheric or at high pressure and temperature conditions. A typical microfluidic setup includes microfluidic chips, chip holder, low-pulse injection pumps, pressure transducer, microscope, and high-performance camera. They may also include inline sensors to detect asphaltene size and concentration. A high-quality digital camera monitors and captures interactions between the in situ phases (oil, asphaltene, and water) and the injected phase (solvent or inhibitor). Quantitative

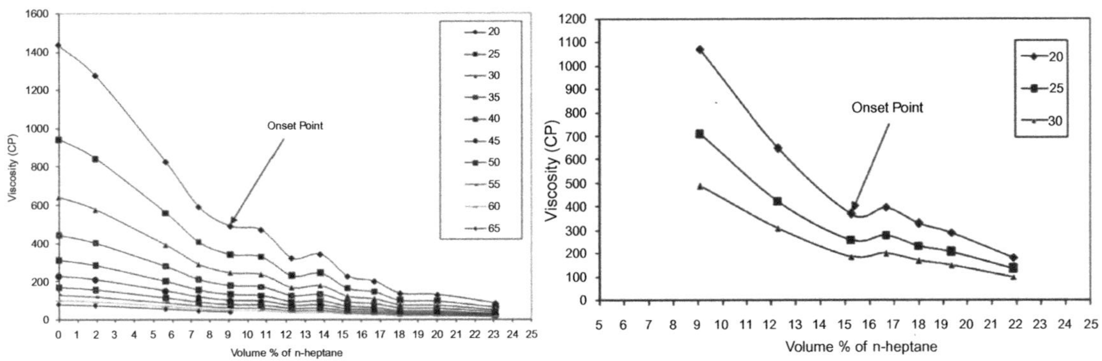

Fig. 5.8 Asphaltene onset point before (left) and after (right) branched DBSA (0.43 wt%) injection in different temperature shown in legend [67].

measurements of the asphaltene deposition can be evaluated using image processing software such as MATLAB, applied on the capture images or videos [75].

Flow loop: Flow loop is another type of test in which the effectiveness of asphaltene inhibitor is evaluated under dynamic flow conditions. In this technique, asphaltene detection is based on either asphaltene deposition inside a pipe [76] or asphaltene capturing through an in-line filter [77]. The changes in the pipeline inside diameter or pressure drop due to the asphaltene deposition with and without the inhibitors indicate the effectiveness of an asphaltene inhibitor [76]. When using the in-line filter in the flow loop, the change in the pressure drop along the filter is considered as a measure of asphaltene deposition severity, or as an indication of the inhibitor effectiveness. Fig. 5.9 depicts a schematic representation of the experimental apparatus of the in-line filter method containing heating and stirring supplements. The vessel contents are circulated in a sintered stainless steel in-line filter, using a metering pump [77].

In each test, a specified amount of oil is prepared at the desired pressure and temperature. Then, the fluid is circulated at a specific flow rate

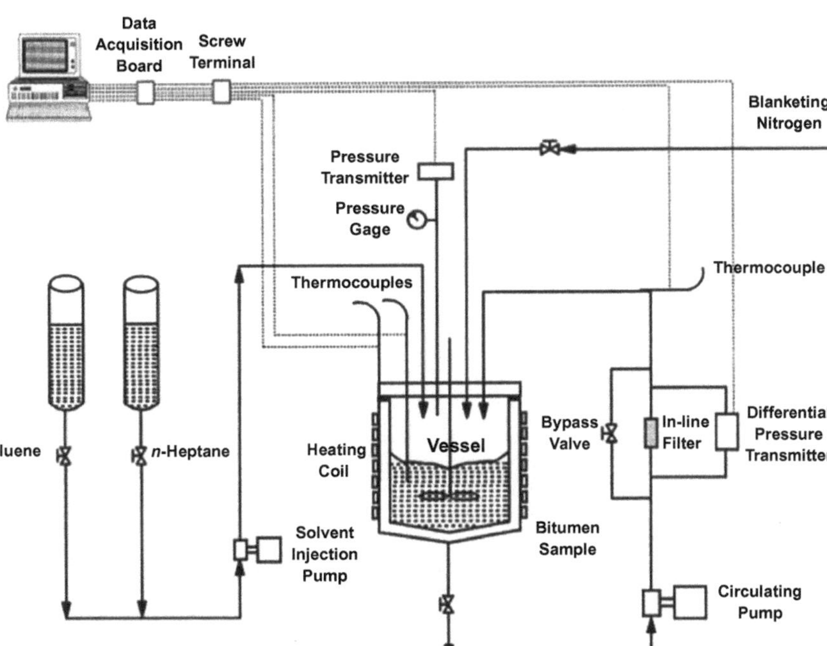

Fig. 5.9 Schematic diagram of the flow loop apparatus [77].

(e.g., 200 mL/min) until the system becomes stable (normally 15–20 min). Next, the precipitant is injected to the vessel at a steady rate (e.g., 2 mL/min). Initially, the pressure drop may decrease due to viscosity reduction, but at a certain concentration of precipitant, the asphaltene particles start to deposit in the pipe; the precipitated asphaltene can be captured by the filter, and therefore increase the pressure drop. The asphaltene precipitation onset point is determined by the ratio of precipitant-to-oil when the pressure drop increases. A similar procedure is applied with various inhibitors to evaluate their effectiveness on the onset point of asphaltene deposition [77].

5.2 Parametric sensitivity analysis

The effectiveness of inhibitors in a hydrocarbon reservoir is related to thermodynamic conditions, the inhibitor type and concentration, oil characteristics, and petrophysical properties. Therefore, the effects of process/operational conditions, fluid properties, and reservoir properties on inhibitor's efficiency are discussed in this section. These parameters are investigated in static and dynamic conditions. Field reports are also discussed to verify the effectiveness of inhibitors in a real situation. Tables 5.1 and 5.2 summarize the research studies which focus on the inhibitor evaluation in static tests for synthetic oil and crude oil, respectively. Also, Tables 5.3 and 5.4 include the research studies which focus on the inhibitor evaluation in dynamic tests for synthetic oil and crude oil, respectively. In the following, the inhibitor evaluation research investigations are discussed in detail.

5.2.1 Effects of process/operational conditions

5.2.1.1 Pressure and temperature, using static tests

The static tests are usually conducted at room pressure and temperature (P/T), and sometimes at elevated P/T (corresponding to the reservoir conditions). The static tests at room P/T are easier and give good preassessments of the inhibitor effectiveness. Although more studies are reported on static tests of asphaltene precipitation assessment at room P/T, there are studies that particularly evaluate effect of P/T on aspahltene inhibitor efficiency. For instance, Alboudwarej et al. [3] and Nassar [78] used the UV–vis method to test the nanoparticle inhibitory effect with synthetic oil in a range of temperatures; they observed that increasing temperature decreases the inhibitors' efficiency. Hosseinpour et al. [15] used HR-TEM and a temperature-programmed oxidation analysis, in addition to the UV–vis method, to investigate acidic (WO_3 and NiO), amphoteric (Fe_2O_3 and

Table 5.1 Results of static tests of asphaltene inhibitor on synthetic oil.

Sol	Prec	Conc	Source (Mwasp)	Additive	Operation T and P	Measurements	Remarks	Ref.
CCl4/CS2	–	–	Manufactured by the Elf Aquitaine	The nanoparticles of (Si, Al, Mg, Zn, Ti)-oxides	Room temp	IR	Physical and chemical asphaltene adsorption with silica and alumina; minor interaction with MgO_2; no interactions with zinc and titanium oxides.	[142]
Not men	*n*C6	5%–15%	Wyoming, California	Dispersants (0–100 ppm), and xylene	–	Dispersion	Up to 87% Asp dispersion with 0–100 ppm dispersants.	[132]
–	*n*C7	60 mg	Fazenda Belem field (3000)	NPH (raw and ethoxylated), D-22, Noram C, *n*-C8, and alkyl benzene	25°C	Spectrophotometer (400 nm)	Long-chain aliphatic alcohols a nd alkyl benzenes are inefficient. Max dispersion (240 mg Asp/m^3) with 1.6% nonyl phenol.	[143]
Tol	*n*C7	1%–2%	Supplied by Mobile research and development (12170)	Phenols and alkyl benzene with different functional groups	25–100°C	Light absorbance (400 nm)	2 wt% inhibitor in nC_7 solubilization: DBSA > NPH > NBDO > NB Phenols with longer alkyl tail without side groups are preferred.	[7,135]
Tol	*n*C5	100 ppm	Lloydminster and West Texas	NPH, steric acid, hexadecylamine, and resin	Room T	UV–vis (400 nm)	With 70% nC_5, the asphaltene dispersity order: hexadecylamine < blank < steric acid < NPH	[144]
Tol	*n*C7	0.0025%–0.0625%	Supplied by mobile research and development (12170)	DP, DPR (0–8 wt%) POM (POM/Asp: 2–8)	295 K	UV–vis (400 nm), FTIR and DLS	Similar solubilization for DP and DPR (15%–100%); DPR is more robust. Increasing POM decreases the Asp size from 500 to 50 nm.	[8]

Continued

Table 5.1 Results of static tests of asphaltene inhibitor on synthetic oil—cont'd

Sol	Prec	Conc	Source (Mwasp)	Additive	Operation T and P	Measurements	Remarks	Ref.
–	*n*C6–12	0.025 g	Supplied by Mobile research and development	DBSA, NPH	6–58°C	UV–vis (400 nm)	Optimum conc. of NPH and DBSA are 7% and 4%, respectively. Less asphaltene dissolves for precipitant with high carbon number.	[83]
–	*n*C7	50 mg	Venezuelan Crude (2100)	Alkyl phenols: SBP, TBP, TOP, NPH, DP, NPE6, and DBSA (1%–16%)	25°C	UV–vis (283 nm), NIR (768 nm)	Inhibitor-asphaltene adsorption: TOP > DBSA > TBP > NPH > SBP > DP > NPE6	[90]
Tol	*n*C7	6 mg/mL	Supplied by PETROBRAS Research Center	Cashew shell liquid, cardanol, poly cardanol, and NPH (0%–2%)	–	UV–vis (780 nm)	100% Asp dispersion with 1% cardanol (at 60% vol/vol nC_7).	[123]
Tol	*n*C7	5 mg/mL,	Venezuelan crude (2100 – 3100)	NPH, DP, NPE6, nonyl phenolic resin, and crude resin (0–0.012 mol/L)	25°C	UV–vis (283–400 nm) Titration, NIR (768 nm)	Plateau in inhibitor concentration is observed. Inhibitor efficiency: NPH resin > n atural resin > NPH	[91, 99]
Tol	*n*C7	0.125%–0.25%,	North Sea	Naphthenic acids, ABSA, hexylamine, octanol, 2-ethylhexanol, inhibitor A, and benzyl alcohol	STD	NIR (1600 nm)	Increasing naphthenic acids conc. improves its inhibitory effect. Hexylamine and 2C4 have three times better efficiency than CNA. ABSA causes more deposition.	[33]
Tol	*n*C7	0.75%	Offshore California, Arab heavy, Safaniya and Argentina	Resin from the same crude	25°C and 80°C	Small-angle neutron scattering	The effectiveness of resin depends on the Asp solution media and the asphaltene polarity and aromaticity.	[119]

Tol	–	0.25–10 g/L	Venezuela, Mexico, France	NPH	303 K	Isothermal titration calorimetry	At higher NPH concentration, a higher number of sites are available for hydrogen bonding and more heat is released.	[120]
Tol	–		Athabasca and Cold Lake (9100–10,400)	Stainless steel (SS), iron, and aluminum powder (0.2–2 g/cm3)	1 atm and 22–60°C	UV–vis (290, 800 nm)	Asp adsorption on: SS > Iron > Al; less adsorption with higher Res and T. More adsorption with Asp from Cold Lake than Athabasca bitumen.	[3]
Tol	–		Puerto Ceiba (1000 – 10,000)	4-NPH, DAIM 2000, and 3000 (0–0.6 g/L)	50°C and 90°C	Vapor pressure osmometry	Optimized concentrations of DAIM 2000, DAIM 3000, and NPH in toluene are 0.1, 0.3, and 0.2 g/L, respectively. The self-assembly of inhibitors reduces their effectiveness.	[137]
Tol	*n*C7	0.5 wt%	Gulf of Mexico, West Africa, North Sea, and Brazil	Kaoline, $CaCO_3$, TiO_2, $BaSO_4$, Fe_3O_4, FeS, SiO_2-i, and SiO_2-o	22°C	UV–vis (336 nm)	Max adsorption 0.26 and 3.78 mg/m^2 for SiO_2-o and SiO_2-i. Increasing the N in Asp increases its adsorption on inorganic particles.	[145]
Tol	–	10–650 ppm	Athabasca vacuum residue	NiO (0–0.5 g)	25°C	UV–vis (410 nm)	Asp adsorption removal increases with concentration of NiO.	[146]
Tol	*n*C7	0.25 g/L	Hassi Messoud and South America	ILs: ILC, ILCA, BMIM (Br-, Cl-), and ILX; Carrying fluid: heptane (1–4 IL/Asp)	–	UV–vis (300 nm), DLS, optical microscopy (×200), IR	Boronic acid improves the inhibitory effect of ionic liquids. [BMIM][Cl], [BMIM][Br], ILC16, and [ILX] disperse 70%–75% of Asp. ILC16 reduces average asphaltene size from 1500 to 50 nm.	[6]

Continued

Table 5.1 Results of static tests of asphaltene inhibitor on synthetic oil—cont'd

Sol	Prec	Conc	Source (Mwasp)	Additive	Operation T and P	Measurements	Remarks	Ref.
Tol	*n*C7	0.1%	Karamay and Lungu	DBSA in heptane (5 g/L) DTAB in ethanol (20 g/L)	25°C	UV–vis (450 nm), ζ-potential (600 nm)	DTAB is more effective with Karamay Asp while DBSA is more effective with Lungu Asp due to their charges.	[92]
Tol	–	10–4000 mg/L	Athabasca bitumen	Alumina	298–328 K	UV–vis (400 nm)	Increasing temp reduces Asp size and self-aggregation. Increasing water to 0.25 vol% decreases Asp adsorption to 45 mg/g.	[78]
Tol	*n*C7	5000 ppm	Middle East	Polyolefin amide alkeneamine (AG), alkylated phenol (4F), and dioctyl sodium sulfosuccinate (AOT)	Room P/T	UV–vis (575 nm), SLS and DLS	1000 ppm 4F and 15 ppm AG, delay onset; AOT is not effective even at 20%. BA and 4F neutralize the Asp charge at CMC.	[14]
Tol	–	1–6 g/L	Puerto Ceiba (Mn = 1000–7000)	4-NPH, DAIM-2000 (M2)	25°C	GPC	0.1 g/L NPH and M2 in 2–6 g/L asphaltene decreased molecular weight. Optimum conc. of NPH and M2 was 0.1 and 0.2 g/L.	[111]
Tol	–	0.1–0.2 g/L	Iranian	Acidic TiO_2 nanofluid (in paraffin)	–	DLS (633 nm)	Acidic TiO_2 decreases the average Asp size from 1890 to 1160 nm.	[98]
Tol	–	100–3000 g/L	Athabasca bitumen	Alumina with different size and acidity	25°C	TGA/DSC	Nanoparticle had higher adsorption capacity: acidic > basic > neutral. Nano-alumina fit better to the Langmuir model, while micro-alumina fit well to the Freundlich model.	[147,148]

Tol	–	150–3000 ppm	Athabasca vacuum residue	Fe_3O_4, Co_3O_4, TiO_2, MgO, CaO, and NiO.	25°C	TGA	Asphaltene, adsorption on CaO > Co_3O_4 > Fe_3O_4 >MgO > NiO > TiO_2.	[118]
Tol	*n*C7		M = 750	BA and DBSA (10–500,000 ppm)	Room P/T	DLS (200–900 nm)	7.5 ppm BA lowers Asp size >6 times; DBSA only effective >2500 ppm BA and DBSA increase the average electrophoretic mobility of fluid.	[20, 22]
Tol	*n*C7	0.01–0.1 *w*/*v*%	Distilled residue of Brazilian crude	Octyl benzoic acid, linear ABSA (10–13 carbon tail)	25°C	UV–vis (850 nm), photon Spec (633 nm)	Asp size increase onset for nC_7 conc: linear ABSA (80%) > octyl benzoic acid (60%) > blank (50%).	[93]
Tol	*n*C7	–	Maya type	C6, C10, C16, N-aryl amino-alcohols with (b) and without (a) boronic acid	Ambient P/T	UV–vis (410 nm), classic turbidimeter	Controversial screening results with light absorbance and Turbiscan tests.	[9]
Tol	–	100–3000 mg/L	Iranian heavy	WO_3, NiO (acidic); Fe_2O_3, ZrO_2 (amphoteric); MgO, $CaCO_3$ (basic)	25–55°C	UV–vis (286 nm), TPO, and HRTEM	Highest and lowest adsorb ratio in equilibrium was for MgO and WO_3.	[15]
Tol	*n*C7	5000 ppm	Alberta crude (729)	OP and DP	Room T	Impedance analysis, HRTEM	5000 ppm OP and DP reduce Asp size by 15% and 20%, respectively. The inhibitors do not affect flocculation onset.	[60]
Tol	*n*C5	1000 ppm	Nigeria heavy crude	BMIM chloride, nitrate, A: 1-methyl-1*H*-imidazol-3-ium-2-carboxybenzoate	25–90°C	UV–vis (346 nm), GC	Inhibitory at 9:1 oil-to-ionic liquid ratio: BMIM chloride (23% less deposition) > BMIM nitrate > A	[82]
Tol	*n*C7	320 mg/L	Iranian	DR (in toluene) (20–50 mg/mL), Toluene and hazelnut oil (50–200 mg)	–	Imaging	Increasing DR conc. decreases average Asp particle size. DR is a more effective inhibitor than hazelnut oil.	[149]

Continued

Table 5.1 Results of static tests of asphaltene inhibitor on synthetic oil—cont'd

Sol	Prec	Conc	Source (Mwasp)	Additive	Operation T and P	Measurements	Remarks	Ref.
Tol	*n*C7	–	Supplied by Mexican Petroleum Enterprise	Zwitterionic liquid	1 atm and 25–90°C	UV–vis (410 nm)	2 g/L ZL has 85% inhibitory effect. Increasing T decreases Asp deposition, but, no change in ZL efficiency.	[4]
Tol	*n*C7, CO_2	–	–	Fe_3O_4 (0%–1%)	<8 MPa <70°C	Microscopy, IFT measurement	0.5% Fe_3O_4 increases onset point from 6.2 to 6.55 MPa. The adsorption capacity of Asp is related to its H/C ratio and N-content.	[45,139]
Tol	*n*C7, CO_2	0.5–1 wt/v%	Brazilian crude and Asphaltic residue	DBSA (0.025–0.25 frac.), Commercial asphaltene stabilizer (CAAS) (0.05–0.1 frac.)	25°C, <6.89 MPa (for CO_2)	UV–vis (650–850 nm)	DBSA is effective without CO_2; more deposition occurs when CO_2 is present. CAAS can be effective with CO_2. CAAS effectiveness decreases by increasing its concentration.	[138]
Tol	*n*C7	1000 mg/L	Colombian crude (1826)	Silica gel nanoparticle; acidic (Ac), basic (B), and neutral (N) (11–240 nm)	Room T	UV–vis (300 nm), DLS	Asp size is 1000–1700 nm (for N) and 11–240 nm (for Ac).	[69]
Tol	CO_2	100–1500 mg/L	Shengli crude	Al_2O_3	60°C <11 MPa	Spectrophotometer (450 nm), IFT test	1 wt% nanoparticle causes 12% more asphaltene stability (onset pressure increases from 7.02 to 7.53 MPa).	[88]
Tol	*n*C7	<500 ppm	Mexican (555)	NPH formaldehyde resins (NPFR), phosphopropoxylated asp (PPO)	30–195°C	Gravimetry	100–1600 ppm NPFR and PPO are effective (up to 74% reduction). Better efficiency with increasing the polymerization degree.	[79, 80]

Tol	*n*C7	–	Iraq	Poly(vinyl toluene-*co*-α-methyl styrene)	STD	Turbiscan	Optimum conc. of 6% inhibitor significantly reduces Asp deposition.	[23]
Tol	*n*C7	5 g/L	Arabic heavy	AMO, AMA, and AMCO (in toluene)	STD	UV–vis (290 nm), Optical microscopy, DLS	AMCO has an optimum concentration (50 ppm), although the inhibitory effect of AMA and AMO increases by increasing concentration. (AMA: the best)	[129]
Tol	*n*C7	0.5 wt%	Canada bitumen	DP, HP, OP, and isoDP	–	DLS	Asp particle size: HP (1.05 μm) > OP (1.05 μm) > DP and isoDP (0.3 μm)	[141]
Tol	*n*C6	100–1000 ppm	Iranian	SiO_2, Al_2O_3, and MgO	25°C	UV–vis (<600 nm)	The adsorption strength order: $SiO_2 > MgO > Al_2O_3$.	[74]
Tol	*n*C7	0.5–5 wt%	Iranian	Fe_3O_4 (amphoteric), NiO and γ-Al_2O_3	Ambient	UV–vis (700 nm)	Inhibitory: γ-Al_2O_3 > NiO > Fe_3O_4	[150]
Tol	*n*C7	1.2 g/L	Athabasca bitumen	DBSA (0–25,000 ppm)	STD	Solubility test	Min conc. of DBSA increases with nC_7 fraction.	[134]
Tol	*n*C7	400 mg/L	Iranian heavy	NiO, Fe_3O_4	Room T	UV–vis, DLS	0.7 mg NiO/mL synthetic and 1 mg Fe_3O_4/mL synthetic increased the onset point 8 and 24 vol% nC_7, respectively. (optimum conc.)	[89]

Table 5.2 Results of static tests of asphaltene inhibitor on Crude.

	Fluids				Operation			
Source	API	Res	Asp	Additive	T and P	Measurements	Remarks	Ref
Bitumen	–	–	16	SiO_2 nanoparticle	Room temp	IR	Physical and chemical adsorption of bitumen with silica.	[142]
Virginia Hills	–	–	–	Alkyl aryl sulfonic acid, polymeric, and commercial	Ambient P/T	Ultrasonic particle analyzer	Limited Asp growth with up to 2800 ppm inhibitors.	[151]
Cold Lake bitumen	10	25	13	Phenanthrene, naphthalene, xylene, toluene, indole, quinoline, DBSA benzothiophene, 2-methoxy-4-methylphenol, tetralin, decalin, and NPH	60–80°C	Thermal	Inhibition efficiency: phenanthrene > toluene > naphthalene > 2-methoxy-4-methylphenol > tetralin > quinoline DBSA and indole were inefficient. Additive solubility parameter did not correlate with the onset point.	[42,152]
Venezuelan crude	22	18	8	Alkyl phenols: SBP, TBP, TOP, NPH, DP, NPE6, and DBSA (1%–16%)	25°C	NIR (768 nm)	The inhibitor strength to change the onset point: TOP > DBSA > DP > NPH > NPE6	[90]
Brazilian crude	–	–	1–7	L31, L64, F108, Renex (100, 200, and 1000), CTAB and SDBS (2 wt%)	Room T	Optical microscopy (×400 – ×1000)	CTAB and R1000 highest increase in onset point (from 3.2 to 3.9 mL nC_7).	[117]
–	–	13	3–4	Commercial inhibitors (50–2000 ppm)	60–140°C <6000 psi	Optical laser light transmission	50 ppm inhibitor is successful in the field. More Asp deposition at lower T due to wax precipitation.	[24]
Brazilian crude	–	–	3–7	Ethoxylated alcohols and phenols; sulfonic and carboxylic acids; and alcohols	STD	UV–vis (400 nm)	Nonionic surfactants (ethoxylated alcohol/phenol) and DBSA are superior with crude oil, and synthetic oil, respectively.	[110]

–	–	–	46–171 ppm	DBSA, NPH, DP, maleic anhydride, and POM (0–1 wt%)	–	FTIR and ASTM D3279	No effect from NPH, DP. Initially 171 ppm Asp; DBSA 118 ppm; Initially 46 ppm Asp; 0.3% POM 164 ppm; 1% polymer 89 ppm; 1% maleic anhydride 1028 ppm.	[121]
–	18–28	3–21	1–9	DBSA	–	NIR (768 nm)	Diluted crude (67%) with #Base = 2–4, #Acid =0.03–0.5 DBSA effectiveness reduces with base number of the oil.	[32]
Kuwait	–	6	3	DO and Res of crude, toluene, NPH, DBSA, and DR (0%–80%)	STD	UV Spectroscopy	Inhibitory: DR > DBSA > NPH > R > toluene > DO. 20% DR is effective; 80% toluene and DO are required to be effective.	[5]
–	22	19	6	DDA, NDP, SM, NPH, NPE6, and DBSA (5–10 v%)	–	NIR (768 nm)	Adsorption energy order: DDA < SM < DBSA < NPE6 < NPH < NDP Onset point of diluted crude increases 5–38 mL nC_7 by reducing adsorption energy.	[153]
Maya, Cold lake, Kuwait, Arab	–	–	–	Inhibitors with COOH, OH, NH_3, and SOOH; (tail/ring variations)	Room T	Toluene equivalence test	The existence of branches prevented crystallization of Asp long carbon tails.	[113]
–	33–35	6–9	1–3	Resins (1.6%–14.7%); DBSA (0.3%–5%)	Ambient	Filtration	Resin with high polarity cause more inhibitory effect. 1% and 5% DBSA change asphaltene deposit from 2.5% to 3% and 2.4%.	[96]

Continued

Table 5.2 Results of static tests of asphaltene inhibitor on Crude—cont'd

Source	Fluids: API	Fluids: Res	Fluids: Asp	Additive	Operation T and P	Measurements	Remarks	Ref
Saskatchewan	28–34	9–17	1–5	DBSA, NPH, and toluene (1000 ppm)	304–338 K 17.2 MPa	Solid detection system	The inhibitory performance of DBSA, NPH, and toluene also depends on the precipitating fluid (CO_2 and nC_7 here).	[125–127]
MuH, Shengli, Xinmin	–	–	1–4	Alkyl phenol and benzene sulfonic, alkyl pyridinium with (Cl, BF_4^-, PF_6^-) and n-butylisoquinolinium chloride (0%–4%)	9.6–15 MPa, 45–69°C	High P/T cell, weighting, *n*C7 titration	4% ABSA and dodecyl pyridinium (BF_4^-, PF_6^-) decreases deposition from 0.75% to 0% and 0.05%. The ionic liquid with chloride and long tail phenols is the most effective.	[154]
Brazilian	–	–	1–4	Ethoxylated phenols, acids, and vegetable oils (0–0.14 mL/g oil)	28°C	Optical microscope,	spectrophotometer (400 nm), and *n*C7 titration	
Mitigation order: coconut essential				oil > palmytic acid > andiroba oil > linoleic acid > ethoxylated phenols > sandaiwood oil > pinewood essential oil > caprylic acid. DBSA is efficient with synthetic oil but not with crude oil.	[53]			
Boscan, Hamaca, Cerro Negro, Furrial	8–22	16–66	10–20	Xylene (20%–80%), Res. I and Res. II of Boscan and Hamaca, and Cerro Negro Resin II (0%–12%)	Ambient condition	Filter drop spreading (visual)	Xylene is only effective for Hamaca crude at high concentration (> 60%). Some of the resins are effective.	[115]

South America	27	–	1.7	Three commercial additives (<200 ppm)	25°C	NIR (1600 nm), FBLR	Particle size and aggregation kinetic does not change near onset point.	[30]
Gulf of Mexico, Europe crudes	25–38	–	–	Polymeric (A) and nonpolymeric amine (B)	–	Colorimeter (570 nm), FTIR and MS	Interaction of basic inhibitor (polymeric) with carboxylic acids make them inefficient inhibitors; amine inhibitor is basic, which is efficient.	[124]
Alaska	27–30	10–17	3–10	DR, DP, DBSA, and 11 proprietary blends (25–10,000 ppm)	25°C 1 atm	UV–vis (740 nm), optical microscopy (×400), turbidity (850 nm), and PSD	At 500 ppm inhibitor, dispersed aggregates larger compared to blank. Turbidity test shows that surfactants are ineffective. Optimum concentrations are 50 (more effective) and 200 ppm (less effective).	[100]
Supplied by Petrobars and Residue of Brazil refinery	–	–	–	Cardanol (CN), PCN and PSS (0–1 w/v%)	Room T	UV–vis (850 nm)	The PCN inhibitory effect relates to conc. and degree of polymerization. PSS is effective at an optimum level of sulfonation and concentration.	[122]
Marrat Jurassic Kuwait	34–39	2–4	0.8–3	Commercial, Diesel, LCO, and HCO (20 wt%)	3000 psig, ambient T	Solid detection (laser)	Optimum conc. is observed with commercial inhibitors, while the effectiveness of noncommercial inhibitors increases with their conc.	[116]
Middle East	–	–	0.3–0.7	polyolefin alkene amine (BA) (0–2 ppm)	Room T	DLS (658 nm)	0.5 and 1 ppm BA delays the onset more than 3 times. BAB and SB, decrease the Asp aggregate size 2–7 times.	[19]

Continued

Table 5.2 Results of static tests of asphaltene inhibitor on Crude—cont'd

Fluids					Operation			
Source	**API**	**Res**	**Asp**	**Additive**	**T and P**	**Measurements**	**Remarks**	**Ref**
Mexican	0.87[a]	–	3.5	Polyolefin amide alkeneamine (AG), alkylated phenol (4F), dioctyl sodium sulfosuccinate (AOT)	Room P/T	DLS (532 nm)	1000 ppm 4F and 15 ppm AG, delay onset; AOT is not effective even at 20%. BA and 4F neutralize the Asp charge at CMC.	[21]
Iranian	19	–	10	TiO_2, SiO_2, and ZrO_2 nanofluid different pH (in paraffin)	–	Polarized microscopy (×500), FTIR	Basic TiO_2, SiO_2, and ZrO_2 are not effective.	[98]
Diluted El Furrial	21	–	8	Ethoxylated nonylphenols, dodecanol and commercial inhibitors (in toluene)	22°C	Turbiscan backscattering	NPE4, NPE5, and dodecanol cause more deposition; dodecanol is only effective at high conc; though 1 wt% can be effective.	[136]
China	–	–	–	SDJ	40–90°C	Turbidity	1 wt% SDJ shows more inhibition efficiency than benzene.	[25]
Gulf of Mexico, Wyoming	27–35	7–12	1–6	A: Nonylphenol formaldehyde resin, B: polyolefin ester (0–500 ppm)	–	Turbiscan dispersions test	A increases the inhibition by 95% and is more effective than B.	[27]
Mexican heavy	12–21	41–49	11–17	Aromatic polyisobutylene succinimides: (1) [–], (2) A carbon between cycles, (3) A hydroxyl group and (4) A boronic acid	25–60°C	UV–vis (410 nm), optical microscope (×100)	Inhibitory: 4 (75%) > 3 (70%) > 1 (60%).	[10]
Nigeria heavy	–	–	–	BMIM chloride, nitrate, A: 1-methyl-1*H*-imidazol-3-ium-2-carboxybenzoate	25–90°C	UV–vis (346 nm), GC	Increasing oil-to-ionic liquids ratio 2:1 to 9:1 reduce deposited asphaltene from 95% to 23%.	[82]

Hossia	0.94[a]	–	12	Poly(dodecylphenol formaldehyde resin) and propoxylated dodecylphenol formaldehyde copolymers	Ambient condition	UV–vis (190–700 nm)	The asphaltene particles interact more strongly with propoxylated than poly(dodecylphenol formaldehyde). The inhibition efficiency increases with Mw and conc. of inhibitor.	[13]
Brazilian	28–29	–	0.3–2.5	DBSA, C0 and C1	Ambient condition	UV – Vis (850 nm) NIR, LS (1550 nm)	C0 and C1 do not affect aggregation size but delay Asp deposition. Inhibitory: DBSA > C0 ~ C1	[29]
Middle East	37–42	–	1–2	DBSA (5%) and BA (50 ppm) Brine = 8w/w% NaCl	–	UV–vis (700 nm), DLS	No effect of brine on Asp deposition is observed without inhibitors or with BA; brine causes more deposition with DBSA.	[94]
Diluted Crude Canada, USA	11–45	–	1–12	DBSA and nonionic polymers in: toluene, paraffinic, olefinic, and aromatic	35°C	Optical scanning, Turbiscan (850 nm)	High Mw polymers are more efficient and robust. The diluted polymer in paraffinic solution is better than that in olefinic one.	[112]
Arvandan field	–	7–9	1–11	DB, DF, FRP, DRO (350–400 ppm)	Lab, Reservoir	Optical microscopy, Turbiscan, SDS	Toluene and xylene are not efficient for dead oil (no onset changes). FRP (best inhibitor) lowers onset P from 4300 to 3600 psi for live oils.	[26]
Iranian	28	16	4	NiO	–	DLS	Nanoparticle decreased the average Asp size from 2190 to 1930 nm.	[71]
North Sea	0.83[a]	–	4	NCAI	25–75°C	ESEM	4 wt% NCAI caused 20 mg/g adsorption on sandstone at 25°C. Adsorption on rock decreased to 15 mg/g by increasing temp.	[155]

Continued

Table 5.2 Results of static tests of asphaltene inhibitor on Crude—cont'd

Fluids					Operation			
Source	API	Res	Asp	Additive	T and P	Measurements	Remarks	Ref
Iranian	30	18	3	Salicylic acid, phthalic acid, NPH, phenanthrene, benzoic acid, and IR95 (0–1000 ppm)	STD and 4500 psig, 242°F	Weight, optical microscopy and PVT cell	Inhibitory:IR95 > salicylic acid > phthalic acid > benzoic acid > NPH > phenanthrene; more deposition reduction at high P/T (50% vs. 44%). Optimum conc.: 250 ppm (IR95, salicylic acid) and 100 ppm for others.	[72]
Iranian	29	–	2	SDES, OP, a commercial inhibitor, lauric acid, and dodecyl amine (<2000 ppm)	Room T	Microscopy (×400) and Zetasizer setup	Inhibitory: Commercial inhibitor (55% size reduction) > OP > SDES > lauric acid > dodecyl amine	[97]
Iranian	34	–	1	SPI (250–2000 ppm)	–	Turbiscan	Optimum conc. of 500 ppm is observed for SPI.	[28]
Middle East	0.76–0.84[a]	5–7	2–3	Commercial inhibitors	1 atm and 25–70°C	UV–Vis, NIR, high-pressure microscopy	At 500 ppm, some inhibitors are effective; their performance reduces with temperature.	[101]
Iranian	–	–	7–10	Henna extract (1–8 g/L)	–	IP 143 method	Optimum effective henna extract is 7 g/L (19% less Asp precipitation).	[73]
Diluted crude	–	–	–	19 commercial inhibitors (6.25–200 ppm); Precipitant: *n*C6	30°C	Asphaltene dispersion test	7 inhibitors had more than 80% efficiency at 200 ppm conc. IB-23 was the best inhibitor which had 70% efficiency with 10 ppm.	[156,157]

Arabic	20	40	11	AMO, AMA, and AMCO (in toluene)	STD	UV–vis (290 nm), O. microscopy, DLS	At 0.2 g/L, dispersion efficiency of Asp: AMA (85%) > AMCO (65%) > AMO (40%); less dispersion effectiveness in synthetic oil.	[129]
Iranian	28	–	–	A mixture of inhibitors (0%–6%)	75°C	Cold finger	The inhibitor at 3% (optimum) had 80% efficiency.	[102]
Iranian	11–19	5–7	9–11	CTAB, SDS, triton X-100, benzene, benzoic acid, salicylic acid, and naphthalene	–	Dispersion test	CTAB (600 ppm) effective for oil A; SDS (300 ppm) effective for oil B due to polarity; salicylic acid > benzoic acid > benzene > naphthalene.	[114]
Iranian	22–27	–	8–11	TiO_2, SiO_2 (In brine: NaCl 30,000–80,000 ppm)	–	NIR (1600 nm)	Optimum condition: conc. =0.04 wt% nanofluid (80% TiO_2, 20% SiO_2); salinity = 4 wt %; and pH = 3.42 and 3.87 for different oils A, B.	[158]
Iranian	–	10	4	DBSA (0–42,000 ppm) Toluene (1–17,200 ppm)	1 atm and high P/T	High-pressure microscopy	Inhibitory: DBSA (2000 ppm optimum) > toluene; both inhibitors are more efficient with the live oil.	[81]
Hassi Messaoud	0.82[a]	4	0.07	Resin and a commercial inhibitor (30–200 ppm)	25–118°C 1–385 bar	Fiber optic laser light transmission	Resin (50 ppm optimum) and commercial inhibitors are only effective at high P/T.	[86]
Live and dead	–	–	–	Inhibitor B to E (33–1000 ppm)	25–80°C <13,000 psi	Analytical centrifuge test, SDS NIR	Inhibition in live and dead oil are similar if T is the same and oil conc. >30%.	[85]

Continued

Table 5.2 Results of static tests of asphaltene inhibitor on Crude—cont'd Fluids

Source	API	Res	Asp	Additive	Operation T and P	Measurements	Remarks	Ref
Iranian	0.82[a]	2	3	New Inhibitor and surfactant (50–250 ppm)	40–110°C	Cold finger, Transmittance	Inhibitors: new (90% at 200 ppm) > surfactant (77% at 250 ppm)	[87]
Kuwait	37	6	0.3–0.8	1E: diesel + fatty acid + diethylenetriamine + kerosene; 1D: polyolefin ester + aromatic solvent; 1 T: naphtha; 1P: ethylene diamine + naphtha; DBSA, salicylic acid (0–1%)	Ambient condition, 60–110°C 9000–800 psig	UV–vis (640 nm), SDS + NIR	Inhibitory efficiency (live and dead oil): 1 *T* > 1E > 1D > 1P > salicylic acid > DBSA	[84]

[a]The values in the API coulmn which are remarked are denisty (g/cm^3).

Table 5.3 Results of dynamic tests of asphaltene inhibitor on synthetic oil.

Sol	Prec	Conc	Source (Mwasp)	Additive	Operation T and P	Media/Measurement	Effect/Remark	Ref
Tol	–	1000 ppm	Northern and southern Italy	A and B: ABSA-based, Additive C and D: complex polymers (2%)	STD	Dolomite Sand pack (k = 2–3 D, q = 15 mL/h)	50 PVI synthetic oil with inhibitors decreases Asp deposition by 7–11 mg.	[140]
Tol	*n*C7	20 g/L	Hassi Messoud and south America	ILs: ILC, ILCA, BMIM (Br-, Cl-), ILX (in *n*C7) (80 g/L)	–	Viscosity test (Shear rate = 260 s^{-1})	The apparent viscosity reduces from 16 to 2 Pa.s with the addition of ILC16, ILC12, and [BMIM][Cl].	[6]
Tol	*n*C7	–	Maya-type crude	C6, C10, C16, N-aryl amino-alcohols with (b) and without (a) boronic acid	Ambient P/T	Rheometer (Shear rate = 0–100 s^{-1})	Although 2b and 3b with 0.1 g/L were not effective, 3b with 2 g/L reduced viscosity about 40cp.	[9]
Tol	*n*C7	–	–	SiO_2, NiO, Fe_3O_4 (0–10,000 ppm)	–	Micromodel (q = 800 $\mu cm^3/min$)	2000 ppm nanofluid increases oil recovery by 38% (with Fe_3O_4) to 53%	[139]

Continued

Table 5.3 Results of dynamic tests of asphaltene inhibitor on synthetic oil—cont'd

Sol	Prec	Conc	Source (Mwasp)	Additive	Operation T and P	Media/Measurement	Effect/Remark	Ref
							(with SiO_2); 5000 ppm is optimum concentration.	
Tol	*n*C7	1000 mg/L	Colombian crude (1826)	Silica gel nanoparticle	358 K, 2500 psi	Sand pack ($q = 0.5$ mL/min, $k = 658$ mD, $\varphi = 14\%$)	S_{or} for blank, damaged and treated sand packs are 21.4%, 25.8% and 16.5%. NP improves oil recovery by 11%.	[69]
Tol	*n*C7	0.5 wt%	Canada bitumen	(*iso*C12, C12, C8, and C6) phenol	23°C	Micromodel ($q = 60$ μL/min, mass flux = 24 $gs^{-1}\ m^{-2}$)	Inhibitors change the deposition rate of synthetic oil from 0.07 to 0.04–0.09%/s. (the most reduction/increment: C12/C8) Inhibitors increases deposition in crude injection. (C12 phenol is the worst)	[141]
Tol	*n*C7	1.2 g/L	Athabasca bitumen	DBSA (0–50,000 ppm)	STD	SS and glass bead packings ($q = 0.6$ g/min, $L = 13.5$ cm)	10,000 ppm DBSA increases deposition rate on SS in 0.65 mass fraction nC7 from 0.004 to 0.0075 $g/cm^2/day$ while in the same condition it decreases deposition rate from 0.0065 to 0.0045 $g/cm^2/day$.	[134]

Tol	*n*C7	400 mg/L	Heavy Iranian	Fe_3O_4	Room T	Couette-Taylor flow assurance apparatus, (Shear rate = 22 s^{-1}, and 278 MPa), DLS	If the asphaltene aggregates are bigger than 50 nm, shear stress increases their size; otherwise shear decreases their size. With NP, Asp aggregates are smaller.	[89]
Xyl	*n*C6	4 g/L	Norwegian Continental Shelf	A fatty alkyl-amine (in xylene) (2 g/L)	20°C	Quartz crystal microbalance with dissipation (Shear rate = 1–100 s^{-1})	Adsorption of esterified asphaltene, whole asphaltene, and inhibitor on stainless steel are 4.2, 3.2, and 0.8 mg/m^2, respectively; xylene wash after inhibitor increases asphaltene deposition. Simultaneous injection of inhibitor and asphaltene decreases the deposition. Using 75% hexane, xylene wash and 2 g/L inhibitor decreases Asp depositions from 19 to 15 and 2 mg/m^2, respectively.	[103]

Table 5.4 Results of dynamic tests of asphaltene inhibitor on crude.

	Fluid properties							
Source	API	Res	Asp	Additive	Operation T and P	Media/Measurement	Effect/Remark	Ref
Alaska and West Texas	–	–	–	Polymeric inhibitor (165–385 gal), activator (160–400 gal)	125 °F, ~3000 psi	Field ($\varphi = 5\%–18\%$)	Xylene temporarily increases production to 350 bbl/d; inhibitors increase Alaska (and Texas) well production to 335 bbl/d for 280 d (11 bbl/d for 90 d)	[104]
Athabasca, Cold Lake	8–11	24–26	15–18	*n*C7–*n*C12 precipitant and four types of solvents and surfactants	180–560 kPa, 60–120°C	Flow loop with filter (60 μm)	DBSA is the best inhibitor with both Cold Lake and Athabasca bitumen. Least efficient: tetralin (for Cold Lake) and naphthalene (for Athabasca).	[77]
Venezuela, the Persian Gulf, the Adriatic Sea and the US Gulf of Mexico	25–38	–	1–10	–	105–290°F 3500–4650 psi	Field ($\varphi = 3\%–11\%$, k = 7 mD)	The inhibitors treat wells and better than ordinary methods.	[133]

China	–	–	–	SDJ	40–90°C	Viscometer	The viscosity of blockage reduces by adding SDJ and increasing T.	[25]
Colombian heavy	19	–	–	Nanofluid alumina in water (Diameter = 90 nm and concentration = 500 mg/L)	323 K and 6.9 MPa	Silica sandpack (30–40 mesh, φ= 33%)	Oil effective permeability fresh sand pack was 521 mD, which increased to 696 mD by NP but that reduced to 390 mD due to inducement. NP injection increased the recovery by 7%.	[70]
Cupiagua Sur crude	38	–	–	Alumina nanofluid	–	Cupiagua Sur core (k = 20 mD, φ=6%)	Nanofluid injection reduces skin damage from 99% to 34% and increases effective oil permeability from 0.1 to 11 mD.	[17]

Continued

Table 5.4 Results of dynamic tests of asphaltene inhibitor on crude—cont'd

Source	Fluid properties: API	Res	Asp	Additive	Operation T and P	Media/Measurement	Effect/Remark	Ref
Iranian	18	13	14	Toluene, linear/branched DBSA, triethanolamine, sodium lauryl ether sulfate, coconut diethanolamide, ethoxylated fatty alcohol	20°C	Viscometer	Sodium lauryl ether sulfate has no inhibitory effect. 1000 ppm coconut diethanolamide is the most effective inhibitor. Linear DBSA is more effective than branched DBSA.	[159]
Supplied by Mexican Petroleum Enterprise	31	20	2	Zwitterionic liquid	90°C, 2400 psi	Cores ($q = 1$ mL/min, $k = 9$–139 mD)	Recovery improvement of 5%–10% (max for Bedford lime) is observed due to the relative permeability and wettability modification	[4]
Mexican heavy, Canadian bitumen,	10–13		14–34	1-(C2–C12)-3-methylimidazolium chloride, 1-butyl-3-methylimidazolium (Cl- or	25–60°C	Rheometer (shear rate = 0.01–100 s^{-1})	Dodecyl pyridinium chloride gives maximum viscosity reduction (35%).	[160]

Venezuelan bitumen				SCN- or BF-), dodecyl (pyridinium or thiazolium) chloride			Increasing the alkyl tail and decreasing the anion radius increase the inhibitor efficiency; imidazolium chlorides with long alkyl tail, however, self-aggregate.	
Iranian	28	16	4	NiO nanoparticle (in CO_2)	120°C, 5400–6600 psi	Carbonated cores (φ=8%–16%, k = 3–7 mD, q = 6–18 mL/h)	Miscible CO_2 injection (P = 6000 psi) reduces permeability and porosity 20% and 1.2%, respectively; using NP coinjection, the damages are 2% and 0.24%. NP coinjection increases recovery by 7%.	[71]

Continued

Table 5.4 Results of dynamic tests of asphaltene inhibitor on crude—cont'd

Source	Fluid properties: API	Res	Asp	Additive	Operation T and P	Media/Measurement	Effect/Remark	Ref
–	0.89[a]	–	8	DBSA (2.5%–5%), Toluene (5%–50%)	Room T	SS capillary tube (L = 30 cm, r = 0.05 cm, q = 0.03–0.7 mL/min)	5% DBSA causes 66% ΔP reduction and 25% toluene causes 50% reduction. Increasing inhibitor concentration increases ΔP reduction.	[76]
Iranian	30	18	2.5	IR95	4500 psig, 242 °F	Carbonate core (k = 0.3 mD, φ = 10%), x-ray spectroscopy	Significant ΔP reduction with inhibitor. IR95 reduces the permeability reduction slope.	[72]
Iranian	34	–	1	SPI	173 bar, 50°C	Sandstone core (φ = 10% and k = 3 mD)	CO_2 injection reduces permeability reduction by asphaltene precipitation. 500 mg/L inhibitor decreases ΔP by 2.5 times.	[28]

Shengli	0.8[a]	29	8	Al_2O_3 nanoparticle (0.1–1 wt %), Brine: NaCl 7500 mg/L	60°C, 2.5–16 MPa	Shengli cores (q = 0.1–2 mL/min, k = 7–10 mD, φ = 10%–15%)	Less permeability reduction with increasing conc. of NP; optimum = 0.5 wt% 1:10 vol NP/vol CO_2 limits permeability reduction to 3%; simultaneous co-inj of NP + CO_2, however, is more efficient than cyclic co-injection.	[88]
Iranian	–	–	7–10	Henna extract (7 g/L)	–	Sandstone outcrop (L = 3.14 in, q = 1 mL/min)	Henna extract injection reduces the permeability reduction by 20%, Asp deposition by 22.5% and ΔP by 24 psi.	[73]

Continued

Table 5.4 Results of dynamic tests of asphaltene inhibitor on crude—cont'd

	Fluid properties							
Source	**API**	**Res**	**Asp**	**Additive**	**Operation T and P**	**Media/Measurement**	**Effect/Remark**	**Ref**
Venezuelan	28	–	8	A polymeric inhibitor	268–300°F, 5600–6839 psi	Field ($\varphi = 10\%$, k = 5–15 md)	100 ppm inhibitor propagate 5 ft. in BOQ-24 and 300 ppm inhibitor squeezed 3 ft. in BOQ-20 caused 400 BOPD more for 5 months	[109]
Arabian Heavy	20	40	11	AMA in toluene (0.5–5 g/L)	10°C	Viscosity (shear rate = 0.01–100 s^{-1})	Increasing the IL concentration and shear rate decrease the crude viscosity	[129]
Iranian	11–19	5–7	9–11	CTAB (in *n*-butanol), SDS (in *n*-butanol + water), Brine = 22% KCl	90°C	Limestone cores (kw = 12–17 mD, $\varphi = 20\%$–22%, L = 7 cm, D = 3.7 cm q = 0.2 mL/min)	CTAB injection reduces oil effective permeability to 2.93 mD which was reduced to 2.56 mD in the absence of inhibitor. SDS injection reduces the oil effective permeability to 4.5 mD, which was reduced to 4.16 mD in the absence of inhibitor.	[114]

Iranian	28.5	–	–	Novel	Reservoir P/T	Carbonate core ($k = 16$ mD, $\varphi = 17\%$, $q = 0.01$–0.1 mL/s)	The inhibitor efficiency increases up to 77% at 3% inhibitor (optimum concentration)	[102]
Iranian	17–18	5–13	14	Toluene, linear and branch DBSA, and cocamide diethanolamine (CDEA)	20°C and atmospheric P	Dynamic viscosity measurement	10,000 ppm CDEA increases onset point up to 11%. Linear DBSA increases onset to 12.4% and branched DBSA is less efficient.	[130]
Iranian	0.82[a]	2	3	Novel and a commercial surfactant (200 ppm)	700–4300 psi	Carbonate core ($k = 16$ mD, $\varphi = 17\%$, $q = 2$ mL/min)	The squeeze lifetime of new inhibitor is 1.86 times more than the commercial one.	[87]
–	0.83–0.88[a]	5–16	1–5	Commercial inhibitors (I8-I9-I15); HP, OP, and DP (60–200 ppm)	25°C	Multisection packed bed with carbon steel spheres	Using inhibitors, the dispersity efficiency is 4%–47%, increasing the deposition by 24%–47% (depending on oil and inhibitor types).	[131]

[a]Density (g/mL).

ZrO_2), and basic (MgO and $CaCO_3$) nanoparticles as asphaltene inhibitors with a synthetic oil. The asphaltene adsorption per specific surface area of the nanoparticles decreased when the temperature was increased; this observation is in agreement with other reports [3, 11, 78]. The asphaltene-nanoparticle adsorption is exothermic which justifies the decreased nanoparticle inhibitory effect with increasing temperature. The effectiveness of polymeric [79, 80], surfactant [72, 81], and ionic liquid [4, 82] inhibitors usually has a direct relationship with temperature and pressure. For instance, Permsukarome et al. [83], Karambeigi et al. [72], and Zahedi et al. [81] showed that increasing P/T increases the effectiveness of inhibitors with varying level of enhancement, depending on the type of inhibitor, the concentration of the inhibitor, and the type of oil (Fig. 5.10) [81].

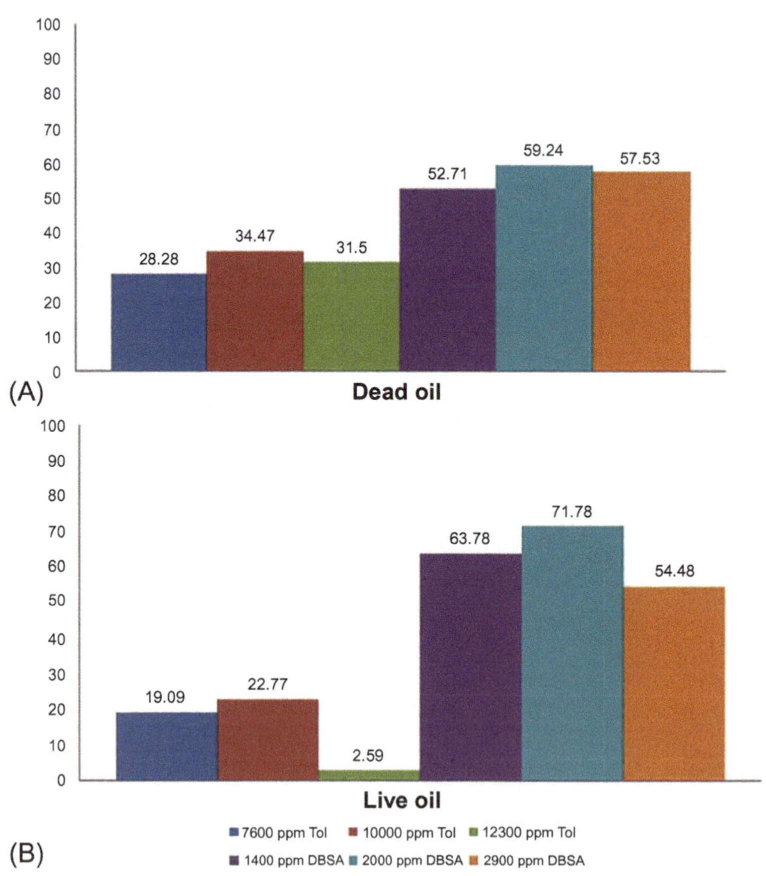

Fig. 5.10 Percentage of asphaltene precipitation reduction in dead (A) and live (B) oil by DBSA addition [81].

Some studies investigated the inhibitors' effectiveness at elevated P/T and compared that with the atmospheric condition [26, 84, 85]. For instance, Ghloum et al. [84] studied several inhibitors (noncommercial and commercial) with dead oil at room P/T and at 1000 psi and 60°C, also with live oil under reservoir conditions (8000 psi and 110°C). Fig. 5.11 shows the inhibitors' efficiency on the oil of the stock tank. The commercial inhibitors had a greater efficiency, and a similar trend was observed at elevated P/T condition compared with room condition based on their study. Firoozinia et al. [26] showed that nonylphenol-formaldehyde resin modified by polyamines (FRP) has superior inhibitory effect compared to polyisobutylene succinimide (DB) under room conditions; a similar trend was observed under reservoir conditions for two different oils.

In some cases, inhibitors lost their efficiency with increasing temperature due to inhibitor deterioration at high temperatures [24, 82]. For instance, 1-methyl-1*H*-imidazol-3-ium-2-carboxybenzoate decomposes at 50°C and loses its efficiency [82]. Ovalles et al. [79] tested the nonylphenol formaldehyde resin (NPFR) [79] and phosphopropoxylated asphaltenes [80] and found that the maximum efficiency of polymeric inhibitors with different polymerization degrees is varied at different temperatures. For example, the maximum reduction of asphaltene deposition with NPFR-4800 (at 12%) and NPFR-900 (at 18%) are at 35°C and 195°C, respectively [79]. Chávez-Miyauchi et al. [10] studied effects of four types of aromatic polyisobutylene succinimides on asphaltene precipitation when the temperature was increased. Three of the inhibitors lost their effectiveness when temperature was increased as they began to behave like high molecular particles in

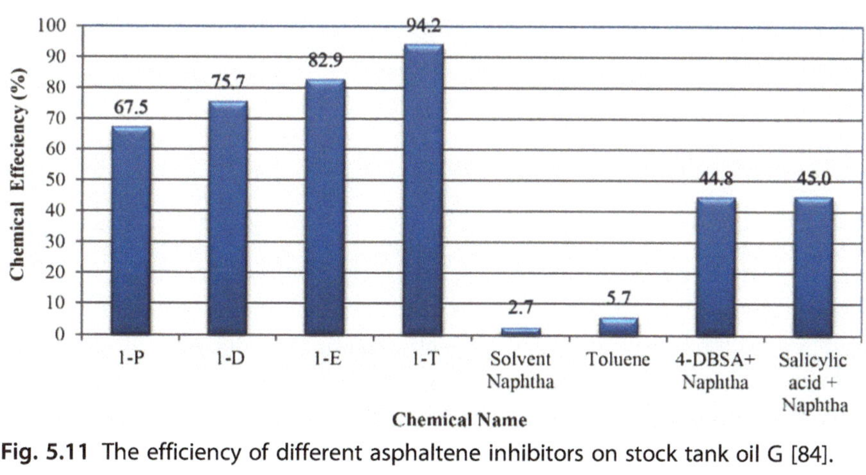

Fig. 5.11 The efficiency of different asphaltene inhibitors on stock tank oil G [84].

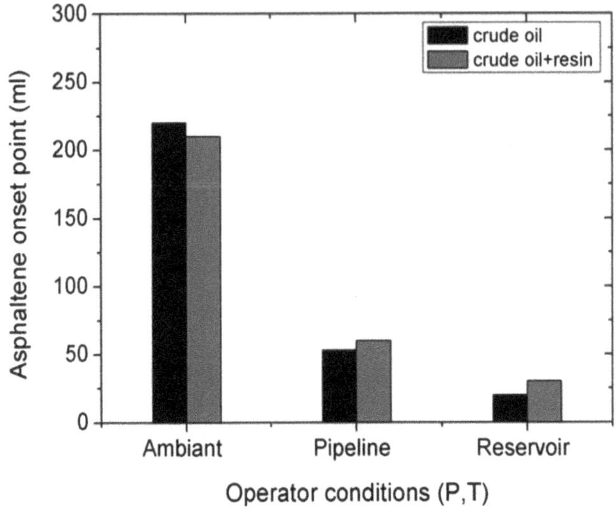

Fig. 5.12 Effect of resins on asphaltene onset flocculation at different operational conditions [86].

the diluted oil, while the fourth one maintained its effectiveness after an increase in temperature [10]. In addition, Gharbi et al. [86] achieved different results by testing both a crude oil resin and a commercial inhibitor under room, pipeline, and reservoir conditions. The resin caused early asphaltene deposition under room conditions, while it postponed the asphaltene deposition at high P/T (Fig. 5.12). The inhibitory effect of the commercial inhibitor with concentration was constant under the reservoir conditions while it had an optimum concentration (at 30 ppm) for maximum efficiency under the room and pipeline conditions [86].

Jennings et al. [85] claimed that if the oil sample contains at least 30% of crude oil, the inhibitor efficiency is comparable with the live oil test if both oil sample and live oil test are conducted at the same temperature. The claim was confirmed by testing four inhibitors (1000 ppm) with an oil sample (at 30% crude) and live oil at 80°C.

5.2.1.2 Pressure and temperature, using dynamic tests

After studying the effects of P/T on asphaltene inhibitors with static tests, their effectiveness should also be evaluated under dynamic conditions. There are not many investigations on pressure effects with the dynamic tests, while the temperature effects are usually studied with the rheology measurements. Although the temperature affects the viscosity, the effectiveness of

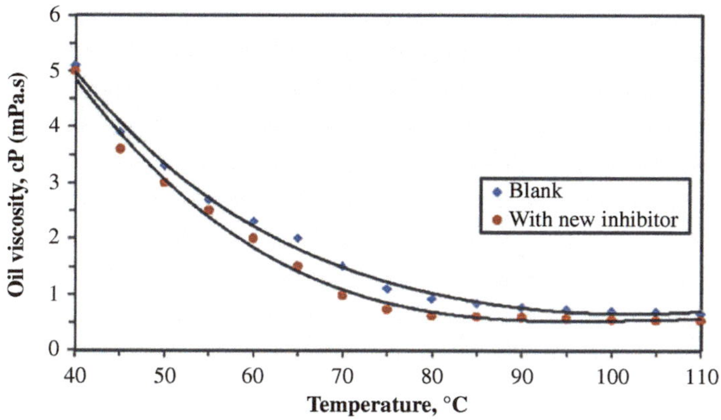

Fig. 5.13 Viscosity changes with temperature: in case of the presence/absence of inhibitors [87].

inhibitors is not significantly influenced. For instance, increasing the temperature from 20°C to 65°C did not change the onset point of the tested oil sample, using toluene, or linear and branched DBSA (as the inhibitors) [67]. In some cases, that viscosity of the oil sample with inhibitor decreased more than the original one when the temperature increased, but not sharply (Fig. 5.13) [87].

5.2.1.3 Static process

Sometimes inhibitors are tested using various methods such as UV–vis, Turbiscan, and other optical techniques to verify the results [8, 9, 14, 21–23, 26, 84, 88–97]. For instance, Mohammadi et al. [98] used FTIR and DLS; both methods confirmed that TiO_2 nanofluids are not effective inhibitors in a basic medium due to the formation of hydrogen bonds, while they are effective in an acidic medium. Hosseinpour et al. [15] used UV–vis and temperature-programmed oxidation (TPO) and studied the efficiency of various nanoparticle inhibitors by reporting the remaining and adsorbed amount of asphaltene. Betancur et al. [69] measured asphaltene adsorption and asphaltene size changes when acidic, basic and neutral silica gel nanoparticle (S11A, S11B, and S11N) were used as inhibitors. Fig. 5.14 shows that the inhibitors had similar results in both tests such that silica gel nanoparticle acidic reduced the asphaltene size more than the others and had the most asphaltene adsorption. In a few cases, the results of various static tests were not conclusive. For instance, Chávez-Miyauchi et al. [10] investigated the effectiveness of four types of aromatic polyisobutylene succinimide

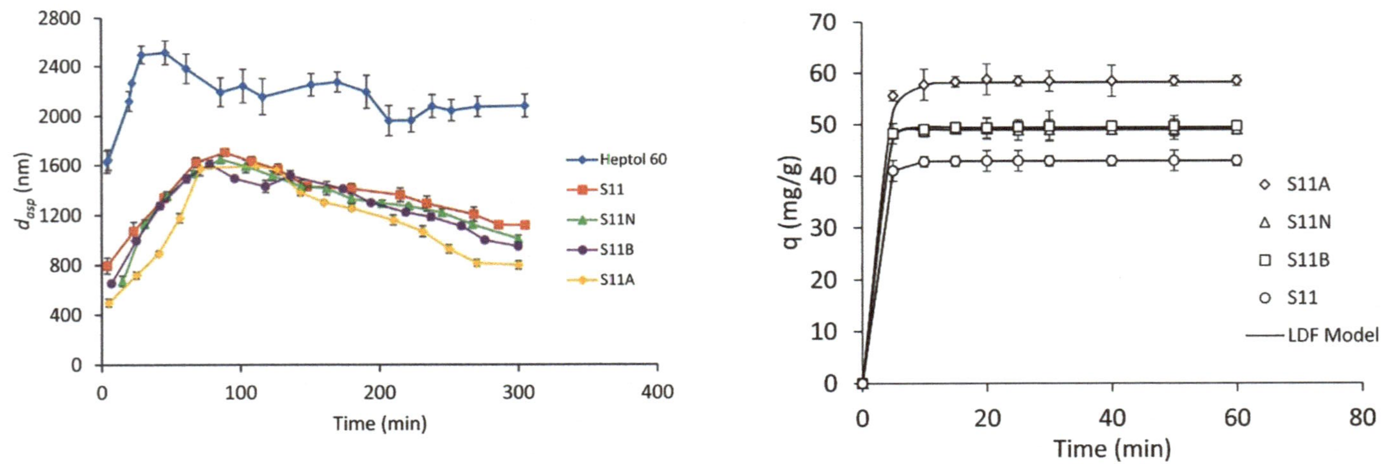

Fig. 5.14 The result of kinetic adsorption of asphaltene (left) and the amount of asphaltene adsorption (right) on silica gel nanoparticle with different sizes and acidities [69].

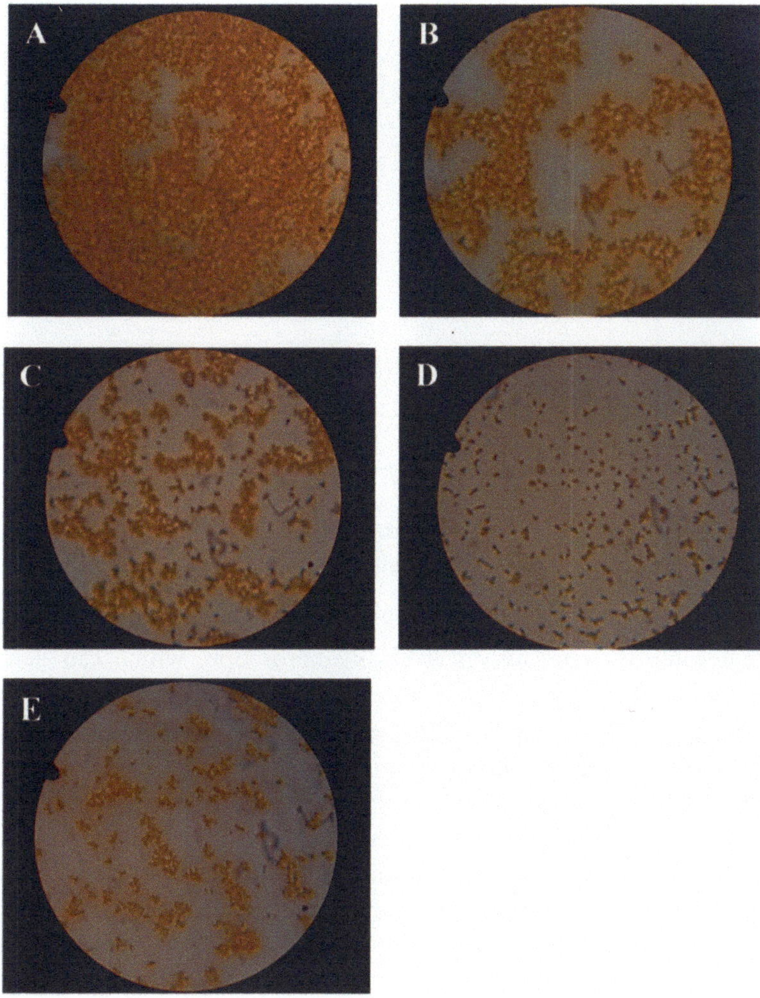

Fig. 5.15 Optical microscopic image with ×100 magnification. (A) without inhibitor, (B) Inhibitor 1, (C) Inhibitor 2, (D) Inhibitor 3, and (E) Inhibitor 4 at 0.2 g/L [10].

inhibitors using UV–vis and optical microscopy in a syntetic oil with an asphaltene extracted from Maya, Mexican crude. As shown in Fig. 5.15, inhibitor 2 did not have an inhibitory effect, but Fig. 5.16 showed the ability of the inhibitor in keeping asphaltene particles dispersed compared to the reference condition [10].

Goual et al. [60] demonstrated that octylphenol and dodecylphenol reduce the asphaltene aggregate diameter and delay the precipitation onset, but they do not affect the flocculation onset. This is due to the use

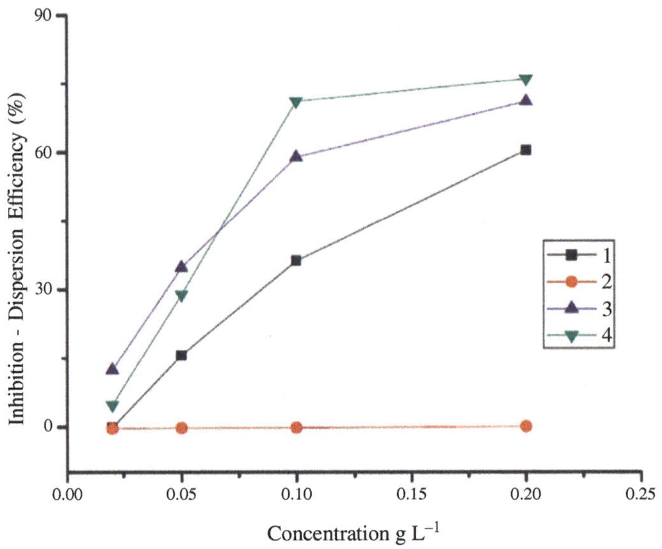

Fig. 5.16 The inhibition efficiency of four inhibitors in different concentrations [10].

of high-resolution transmission electron microscope, which can detect a particle size of 50–100 nm (that is relevant to the flocculation onset), while other optical methods' detection limit is 1 μm (which is relevant to the precipitation onset) [60]. Boukherissa et al. [6] used UV–vis to measure the asphaltene flocculation onset point and flocculation kinetics, and DLS to measure the asphaltene aggregate size. Although the flocculation onset point did not change much after the addition of ionic liquid, the size and kinetic flocculation tests exhibited similar inhibitory effects with ionic liquids [6]. Leon et al. [99] found that although the ultimate adsorptions of nonylphenol and nonylphenol resin on asphaltene are the same, the adsorption of the nonylphenol resin on asphaltene is slightly faster [99]. Marugan et al. [30] investigated the precipitation onset point, the kinetic of asphaltene precipitation, and asphaltene particle sizes of a crude with three commercial inhibitors. The onset point analysis showed that inhibitor A is the most effective inhibitor (Fig. 5.17), while the particle size distribution and kinetic aggregation tests showed that Inhibitor B is the most effective inhibitor (Fig. 5.18) [30].

Kraiwattanawong et al. [100] tested different inhibitors using UV–vis spectroscopy, optical microscopy, turbidity, and PSD. The UV–vis spectroscopy showed that the onset point of samples increases with the addition of 500 ppm inhibitors. The optical microscopy showed that the inhibitor

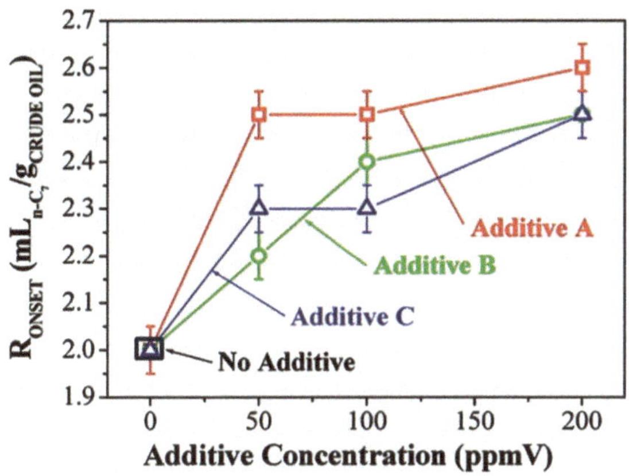

Fig. 5.17 The effect of inhibitor concentration on the onset point of the D-190^{+} residue [30].

keeps asphaltene aggregates dispersed in a smaller size, but it could not stop asphaltene aggregation growth. In addition, unlike the UV–vis test, the turbidity and PSD tests considered DR, DP, and DBSA to be ineffective inhibitors [100]. Juyal et al. [27] studied the asphaltene adsorption reversibility by adding inhibitors through standard ADT, modified ADT, and Turbiscan tests. It was found that the ADT test and Turbiscan test verify each other [27]. Nevertheless, the ADT test showed 100% inhibition for an inhibitor in the crude, while the modified ADT showed the efficiency improvement of the inhibitor up to 95%, 30 min after adding the precipitant. After 24 h, the inhibitory efficiency decreased to 85% and then decreased to below 80% after 7 days. These results verified the Kraiwattanawong et al. [100] claim that the asphaltene aggregation does not stop even in the presence of inhibitors, but the rate of aggregation can decrease. Melendez-Alvarez et al. [101] investigated the efficiency of commercial inhibitors in two different procedures, direct and indirect,[a] using NIR spectroscopy and digital microscopy. In the direct procedure [101], the measurement of asphaltene aggregation over time showed that the inhibitor does not stop aggregation, but it reduces the rate of asphaltene aggregation, which is in agreement with the results of Kraiwattanawong et al. [100] and Juyal et al. [27]. In the direct procedure, it was claimed that asphaltene dispersants could shift the onset precipitation of asphaltene. However, the results of the indirect procedure showed that, due to measuring particles at nanoscale, inhibitors neither shift the actual onset nor reduced the amount of asphaltene precipitation [101]. Therefore,

[a] Please refer to Ref. [161] for more information about direct and indirect procedures.

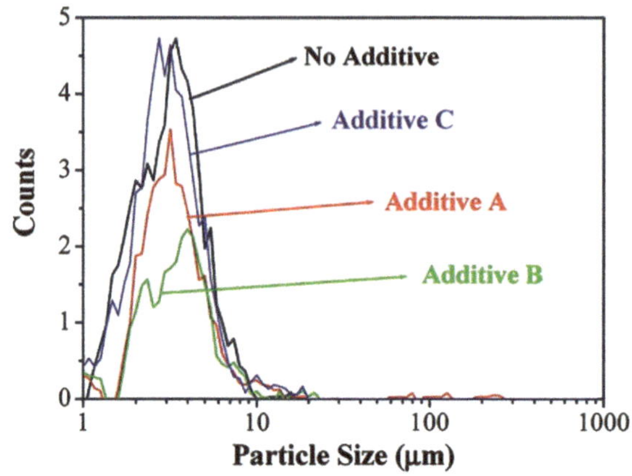

Fig. 5.18 Effect of inhibitor (100 ppm) addition on the particle size distribution of asphaltene [30].

controversial findings about the ability of inhibitors to change the precipitation onset point might be due to the combination of the incompetency of the measurement device (to capture molecular-scale behavior) or that of the test procedure [6, 27, 30, 60, 99, 100].

5.2.1.4 Dynamic process

Dynamic process variables such as flow rate, fluid pattern, and injection pressure conditions can affect the inhibition behavior of inhibitors. Hashmi and Firoozabadi [76] used the flow loop method and studied the flow rate of DBSA and toluene on the dissolution of deposited asphaltene in stainless steel capillary tubes, and used injection pressure as a measure of blockage. The most effective asphaltene removal was achieved at a low flow rate due to the longer residence time. Unexpectedly, applying a high shear rate, using a high flow rate did not enhance the asphaltene removal [76]. Khormali [102] performed a coreflood test and concluded that increasing the hydrocarbon flow rate increases the asphaltene deposition in a carbonate core. Subramanian et al. [103] and Lu et al. [88] studied the effect of inhibitor injection pattern of alkylamine inhibitors and aluminum oxide nanofluid on their inhibitory effects, respectively. They concluded that the simultaneous injection of inhibitors with precipitant and asphaltenic fluid causes less asphaltene deposition compared to subsequent fluid injection [88,103]. Lu et al. [88] stated that the simultaneous injection of nanofluid and carbon dioxide is more effective than cyclic injection scheme. In addition, in cyclic

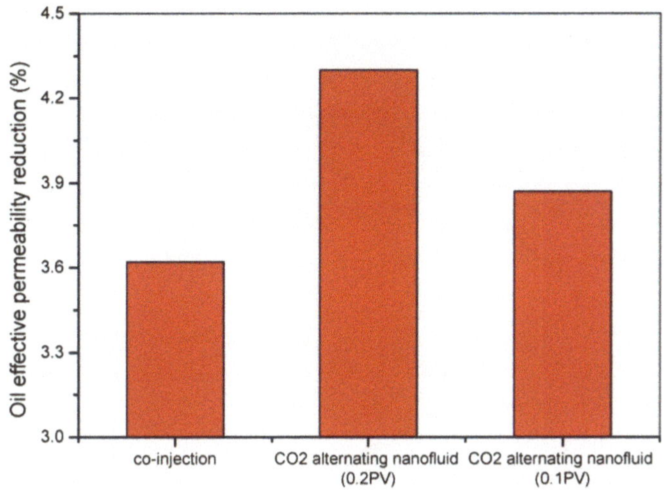

Fig. 5.19 Effect of injection pattern on oil effective permeability reduction [88].

injection, using more frequent cycles with smaller slug size is more effective than less frequent cycles with larger slug sizes (Fig. 5.19).

Lu et al. [88] conducted injection pressure sensitivity analysis and observed that increasing the oil injection pressure above the bubble point causes less asphaltene deposition and permeability reduction, leading to a higher total oil recovery both in the presence and absence of nanofluid flooding (Fig. 5.20).

There are also few asphaltene inhibition studies conducted at the field scale. Allenson et al. [104] proposed using activators before cyclic injection inhibitor and soaking polymeric inhibitors near wellbores to improve the inhibitor adsorption onto reservoir rock. They discovered that the activator decreases the repetition cyclic injection of inhibitors [104]. Yin et al. [105] applied two inhibitors to retard the asphaltene deposition in a West Texas well. The asphaltene was deposited due to CO_2 injection for EOR purposes. The inhibitor delayed the filter to plug four times more than the untreated oil production. Means [106] monitored asphaltene deposition during the treatment with RE-4300 inhibitor in field scale. Means observed a higher concentration of asphaltene stable in the oil in the higher concentration of RE-4300. Al-Qabandi and Al-Naqi [107] studied the inhibition of asphaltene aggregation in Minghashi field deep carbonate reservoir. The inhibitor, called DART, was injected followed by toluene and diesel, and allowed to soak for 3 weeks. The well monitoring after treatment showed that the asphaltene deposition happens but with a significantly lower rate.

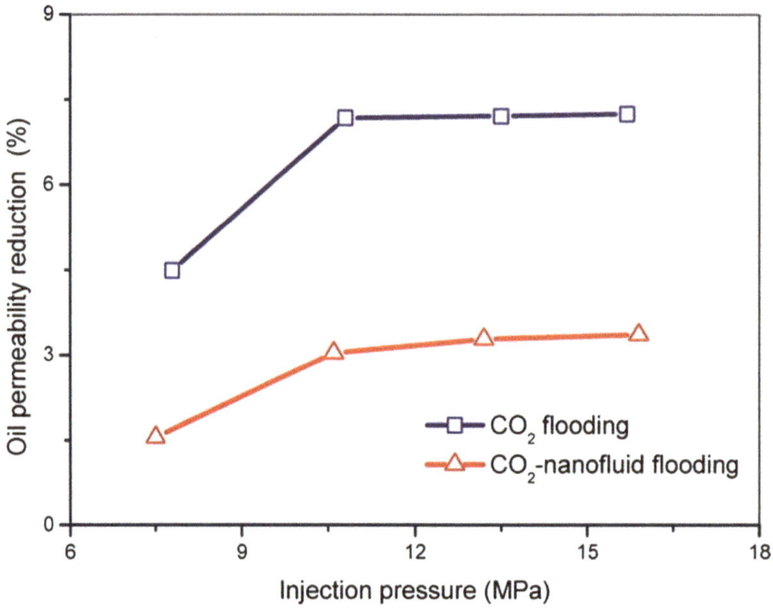

Fig. 5.20 Oil effective permeability reduction in different injection pressure conditions [88].

Del Carmen Garcia et al. [43] studied the asphaltene deposition control in North Monagas Oil Field in Venezuela. They used two commercial inhibitors in five wells with the squeezed method for 48 h soaking time . For most of the cases, the inhibitor was successful in reducing asphaltene deposition. For instance, the production was unchanged in well S9—C for 8 months after the treatment, which also shows the great economic potential of treatment. Sanada and Miyagawa [108] reported a case study of a well with a light crude oil in a Japanese oil field, which was suffering from asphaltene deposition. A chemical was injected near wellbore with a radius of 1.2 m using a coiled tube after a xylene wash. The comparison of wellhead pressure and production rate showed that the chemical treatment is successful. Villard et al. [109] introduced a new generation of asphaltene inhibitors, which can adsorb onto the surface of reservoir rock and reduce the number of cycles of inhibitors' injection. They utilized a synthetic inhibitor in the Boqueron field of Venezuela and observed a delay in the demand for asphaltene treatment after 60 days of oil production to 180 days. Romero et al. [17] reported that the CPSXL4 well of the Cupiague Sur oil field treatment with alumina nanofluid showed production enhancement from 1700 to 3500 bbl, which remained effective for 8 months (Fig. 5.21). Igder et al. [89] designed a pilot

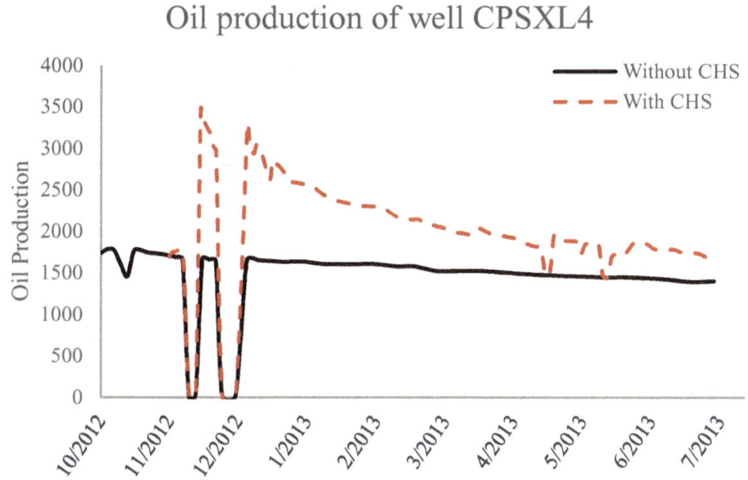

Fig. 5.21 Incremental production after nanofluid treatment in CPSXL4 well [17].

to simulate the near wellbore condition and test the effect of Fe_3O_4 as an asphaltene aggregation inhibitor. They considered 278 mPa as shear stress to have a remarkable asphaltene growth. The aggregation growth was predominant when the aggregates were smaller than 50 nm while the fragmentation became dominant for aggregates larger than 50 nm for both presence and absence of nanoparticles. However, the interaction of asphaltene with nanoparticle surface decreased the asphaltene aggregation [89].

5.2.2 Effects of fluid properties

The properties of the fluids used in asphaltene inhibition analysis—including the characteristics of oils and asphaltenes, the types and concentrations of precipitants, solvents, and inhibitors—affect the efficiency of inhibitors on asphaltene deposition. In this section, we discuss the proficiency of the inhibitors in synthetic oils with different levels of asphaltene [19,20,67,79,91,110–112] as well as crude oils from different sources [10,26,110,113–116], under both dead and live conditions.

5.2.2.1 Concentration and type of inhibitors

The concentration and type of asphaltene inhibitors control the efficiency of asphaltene deposition inhibition. There is no conclusive study to generalize the screening of asphaltene inhibitors for different crude oils. The majority of studies use limited types (and concentrations) of inhibitors for given crude oil samples, and under a limited range of operating conditions. The effect of concentration and type of inhibitors on asphaltene deposition was discussed in detail in Chapter 4; the effect of the asphaltenic fluid is discussed in the following sections of this chapter.

5.2.2.2 Type and concentration of asphaltene in synthetic oil

Synthetic oils are prepared by dissolution of extracted asphaltenes from oil in toluene or other solvents. The behavior of the asphaltene type (at molecular level) is related to the type of interactions between the asphaltenes and inhibitors. The concentration of asphaltene in synthetic oils changes the extent of asphaltene deposition. Asphaltene can be extracted from various oils, using different precipitants and extraction processes. Since asphaltene is composed of a range of molecules, extracted asphaltenes from a particular crude oil using different precipitants can have different characteristics. Alboudwarej et al. [3] showed that asphaltenes from different crudes have distinct amounts of adsorption on metal nanoparticles (Fig. 5.22). Kazemzadeh et al. [45] studied the adsorption of two different asphaltenes on Fe_2O_3 nanoparticles in the presence of CO_2 as a precipitant and observed that the asphaltenes with a lower hydrogen/carbon ratio and those with a higher nitrogen content have higher adsorption affinities. Wang et al. [92] found that inhibitors with a positive charge can disperse negatively charged asphaltenes and vice versa. This was verified by measuring the size of two different types of asphaltenes from different sources using the DBSA and dodecyl trimethyl ammonium bromide (DTAB) inhibitors [92]. The results from Wang et al. [92] research can describe the inefficiency of sodium dodecylbenzene sulfonate (SDBS) in a study by Mohamed et al. [117]. High concentration

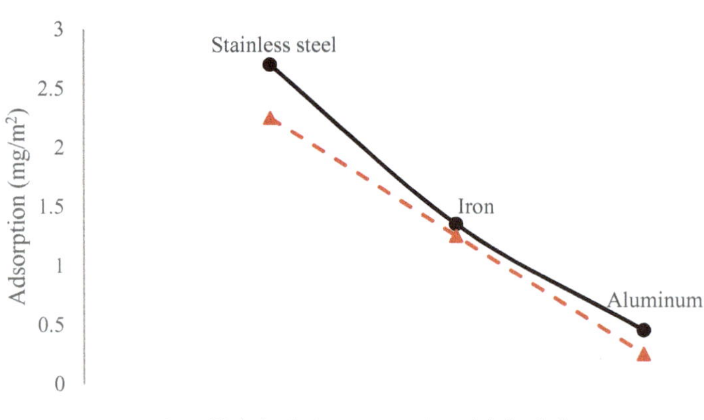

Fig. 5.22 The asphaltene-nanoparticle adsorption for different asphaltene types [3].

zwitterionic liquids (ZL) had a similar efficiency with the asphaltenes extracted from both heavy and light crude oils. At low concentration of ZL, the inhibition efficiency of the case with asphaltene extracted from the lighter crude oil reached a maximum faster than others [4].

Nanoparticles (e.g., NiO) [118] and surfactants [110] showed better performances when the asphaltene was extracted by heptane than that with pentane; the higher polarity and aromaticity of the asphaltene by heptane are among the reasons for such a behavior. Spiecker et al. [119] used resins to inhibit three types of asphaltenes, including whole asphaltenes (deposited asphaltene from heptol containing 50% toluene), soluble asphaltenes (deposited asphaltene from heptol containing 30%–40% toluene), and precipitated asphaltenes (deposited asphaltene from heptol containing 60%–70% toluene) from different oils. Fig. 5.23 demonstrates that different resins may have similar inhibitory effects for all types and sources of asphaltenes [119], while Goual and Firoozabadi [96] stated that various resins cause different levels of inhibition for asphaltene from different crude oils.

In 2018, Subramanian et al. [95] investigated the inhibitory performance of fatty alkylamine with four types of crude oil asphaltenes. The deposition trend of the whole asphaltene, the asphaltene extracted with *n*-hexane (*n*C6), fluctuated after increasing the concentration of fatty alkylamine. The irreversibly adsorbed asphaltene, the most strongly adsorbed portion of the whole asphaltene, was found to have higher net interaction enthalpy with inhibitors compared to the whole asphaltene. However, 3.5 V asphaltene, the first deposited fraction of asphaltene with 3.5 vol *n*C6/g oil, had more net-interaction enthalpy beyond a specific inhibitor-asphaltene mole ratio [95]. The asphaltene inhibitor molecules did not interact with each other through acid-base interactions, and the inhibitor efficiency was similar for the esterified asphaltene with methanol and whole asphaltenes [95]. Although Subramanian et al. [95] stated that acid-base interaction is not the main reason for the inhibition of asphaltene deposition, Hashmi et al. [22] claimed that the acid-base interaction is the main inhibitory mechanism, using DBSA as an inhibitor. In another study, Subramanian et al. [103] investigated the adsorption of esterified and whole asphaltenes on the surface of a stainless steel coated with quartz crystal, using quartz crystal microbalance with dissipation. The synthetic oil passes through the quartz crystals and adsorbs onto the surface. The quartz crystals vibrate and the vibration frequency changes with the mass of deposition. Although the esterified asphaltene showed more adsorption compared to the whole asphaltene, both asphaltenes were removed to the same extent

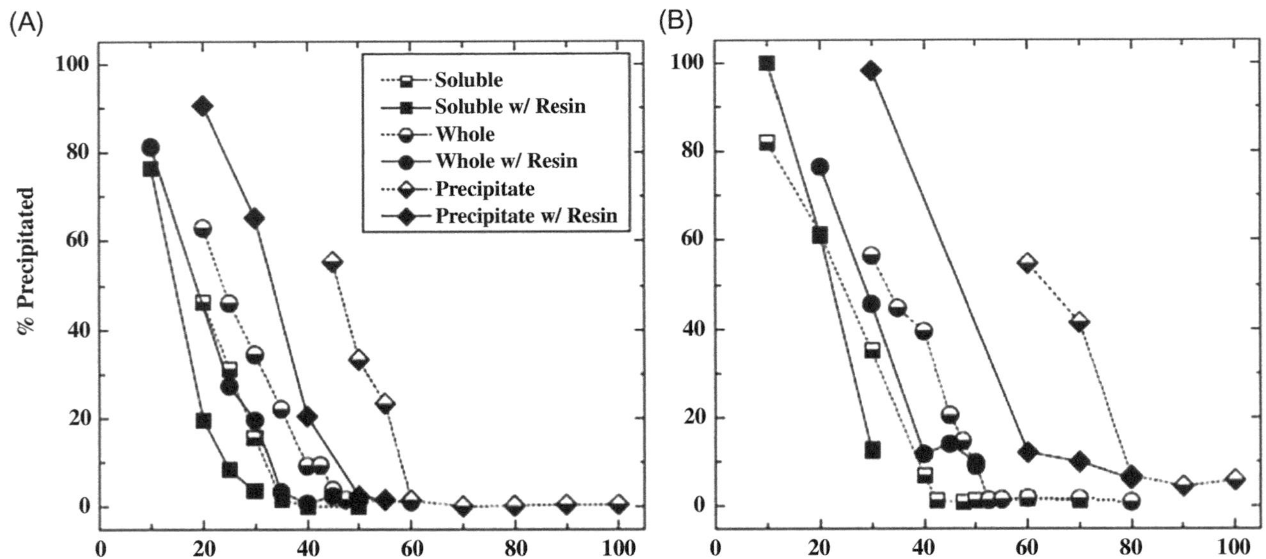

Fig. 5.23 Effect of resin on the precipitation of different types of asphaltenes from oil (A) AH and (B) oil B6. The half-filled mark denotes system without resins [119].

after washing with xylene [103]. Peramanu et al. [77] used a flow loop system and showed that the asphaltenes from Athabasca bitumen have a lower sensitivity for inhibitors with single ring aromatics and hydrogen donor inhibitors compared to the asphaltene from Cold Lake bitumen; both asphaltenes had the same behavior with inhibitors that include heteroatom and surfactant inhibitors [77]. Therefore, the source and type of asphaltenes are important parameters in screening the inhibitor(s).

Besides the types of asphaltenes, the asphaltene concentration in the fluid is another variable affecting the inhibitor efficiency. The asphaltene concentration directly affects the adsorption on metal/metal oxide nanoparticles with/without coating [11, 17, 70, 88]. For instance, increasing the asphaltene concentration from 100 to 1000 mg/L increases asphaltene adsorption on an alumina nanoparticle from 8 to 65 mg/g [78]. Similar behavior was observed with CaO, Co_3O_4, Fe_3O_4, MgO, NiO, and TiO_2 nanoparticles [118]. Merino-Garcia and Anderson [120] investigated asphaltene-nonylphenol (NPH) inhibitor interaction in a range of asphaltene concentrations. For low asphaltene concentrations, the main mechanism was the dissociation of NPH self-aggregation, while asphaltene-NPH interaction became dominant when the asphaltene concentration increased from 0% to 10% [120].

5.2.2.3 Type of oil

The composition of oil containing asphaltenes also affects the inhibitor efficiency during asphaltene deposition inhibition. Pillon [121] studied the effectiveness of poly(maleic anhydride-1-octadecene) with two different oils. In one oil, the asphaltene deposition decreased when the inhibitor concentration was increased, while in another crude, the amount of asphaltene deposition increased at low inhibitor dosages and reduced at above a threshold concentration (0.3 wt% polymer) [121]. These trends were also observed for polycardanol (PCN) and sulfonated polystyrene (PSS) as the asphaltene inhibitors [122] using heptane and acetone as precipitants, respectively [122]. Based on the literature studies, a minimum threshold concentration is required for the polymeric inhibitors to be effective; this threshold is lower for polymers with higher molar weights [8,123]. Madhi et al. [114] concluded that, besides the composition of oil, polarity (such as asphaltene's charge) also affects the inhibitor's performance. Rogel et al. [32] showed that low concentrations of DBSA react with basic components in the crude and produce salt instead of inhibiting the asphaltene deposition; however, at the high concentrations, the inhibitors could neutralize basic components and stabilize the asphaltene. This claim was confirmed by testing DBSA with diluted crude oils with different basicities, and the least effective inhibition was observed for the most basic crude oil [32]. Hashmi et al. [22] also verified that the effectiveness of DBSA is ceased after the addition of triethylamine (basic component) with three different crude oils. Smith et al. [124] found that the oils, containing carboxylic acid, deactivate basic inhibitors, while the protonic inhibitors are effective in stabilizing the asphaltene. Ibrahim and Idem [125–127] studied the effect of asphaltene content on asphaltene precipitation rate, using different inhibitors and precipitators (CO_2 and heptane), and oils with different characteristics and aromaticities. They showed that the rate of asphaltene precipitation is affected by the composition of the asphaltenes such as the presence of heteroatoms (nitrogen, sulfur, and oxygen), the aromatic carbon fraction, and the branching degree of the asphaltene side chains [125–127]. This trend is verified by studying the effectiveness of a few surfactants and commercial inhibitors with different crude oils [27,100,128]. During the evaluation of the inhibitors, either with the synthetic or crude oils, some showed a higher efficiency with crude oils such as ethoxylated alkylphenols, while others showed a higher efficiency with the synthetic oils, as did DBSA [110]. The inhibitors usually behave more effectively in synthetic oils, which is due to fewer intruder components [29,129].

In dynamic asphaltene inhibition tests, the type of crude oils affects the inhibitory efficiency differently, which is verified by experimental [114,130,131] and field [132,133] studies. For instance, Schantz and Stephenson [132] reported different degrees of effectiveness for the polymeric inhibitors in different wells, including those in Bighorn Basin, WY, Green River Field, ND, and Gillette, WY. In addition, Cenegy [133] reported inhibitor effectiveness in wells of the Adriatic Sea and the Gulf of Mexico and wellbores of eastern Venezuela and the south Persian Gulf [133].

5.2.2.4 Type of precipitant and solvent

Researchers have studied the effects of the type and concentration of precipitants and solvents on the effectiveness of inhibitors in asphaltene deposition tests. For metal and metal oxide nanoparticle inhibitors, increasing the concentration of the precipitant increases the asphaltene-nanoparticle adsorption, which implies a greater inhibitory effect [3, 78]. When using surfactants, such as DBSA, the inhibitor's threshold concentration to inhibit asphaltene deposition increased after increasing the nC_7 fraction in the synthetic oil [134]. However, Hashmi and Firoozabadi [20] hypothesized that increasing nC_7 reduces the volume fraction of the asphaltene, which decreases the probability of asphaltene's collisions (and their aggregation) and consequently, decreases the chance of deposition. Melendez-Alvarez et al. [101] stated that the addition of precipitants in the dark fluid could dilute the sample, creating more transparency, which may be mistakenly interpreted by asphaltene precipitation and deposition [101] even though there was no actual deposition. Therefore, the addition of precipitants may lead to incorrect conclusions about the asphaltene precipitation and deposition. The type of precipitant(s) used in the deposition tests affects the efficiency of the inhibitors. For instance, 24% NPH is needed to stabilize asphaltene in synthetic oil when pentane is used as a precipitant [7,135] while 7% NPH is enough to keep the asphaltene suspended when nC_7 is used as a precipitant. The necessary concentration of NPH to stabilize asphaltene will grow to 15% gradually if the used precipitant is changed to heavier normal alkanes from *n*-heptane to *n*-hexadecane. Unlike NPH which has extermum, the required amount of DBSA to inhibit the same asphaltenic fluid increases steadily when changing the precipitant from pentane to hexadecane [7,135]. Permsukarome et al. [83] proposed that the precipitation rate increases in lighter precipitants due to faster dynamics of the intermolecular interactions. They tested precipitants from *n*-hexane to *n*-hexadecane, and showed the fastest kinetics of interactions between the asphaltenes and precipitants occurrs with hexane. Pereira et al. [136] tested the deposition time for cyclohexane, *n*-pentane, and *n*-heptane. The transmittance, which is the deposition rate index, reached 60% for cyclohexane after 50 h, but for *n*-pentane and *n*-heptane, it gradually increased to 50% and 80%, respectively, after 300 h [136]. Barcenas et al. [137] investigated the effect of solvent polarity, and observed that both low- and high-polarity solvents could have disadvantages. In a polar solvent (ODCB), the self-aggregation of the inhibitor was high, causing larger agglomerates. In addition, in a low-polarity solvent (toluene), the polar heads of the inhibitors are blocked, that decreases the inhibitory effect [137]. Ibrahim and Idem [125–127] and Loureiro et al. [138] both compared the CO_2 and nC_7 as precipitants in the presence of DBSA, NPH, as well as a commercial inhibitor, and concluded that the performance of inhibitors is affected by the type of the precipitant used. For instance, DBSA was effective in keeping an asphaltene dispersed in a synthetic oil with nC_7, serving as a precipitant; DBSA was, however, ineffective when CO_2 was used as the precipitant [138]. The commercial inhibitor was effective with both nC_7 and CO_2 [138].

The asphaltene medium also affects the efficiency of the inhibitors in the dynamic inhibition tests. Kazemzadeh et al. [139] investigated the effect of nanoparticles in a micromodel. Once the model was saturated with asphaltene/toluene, water flooding recovered 28% of the oil while SiO_2 nanofluid injection recovered 38% of the oil from the system. When the micromodel was saturated with a mixture of asphaltene, heptane, and toluene, water flooding and SiO_2 nanofluid injection recovered 30% and 50% of the oil, respectively, as shown in Fig. 5.24 [139].

Allenson et al. [104] reported near wellbore injection of a polymeric inhibitor for two wells that suffered from asphaltene deposition upon CO_2 injection and high pressure drop. The inhibitor efficiency lasted for 80 and 90 days in two oil wells and the second time treatment lasted even longer in the case of the well with the CO_2 injection (Fig. 5.25) [104].

5.2.2.5 Effect of water or brine

Although the studies on the effect of brine (or water) on the performance of inhibitors are limited, it appears that brine can have significant influences on inhibitor performance. For example, Nassar [78] tested the asphaltene-γ-Al_2O_3 nanoparticle in synthetic oil and showed that increasing water

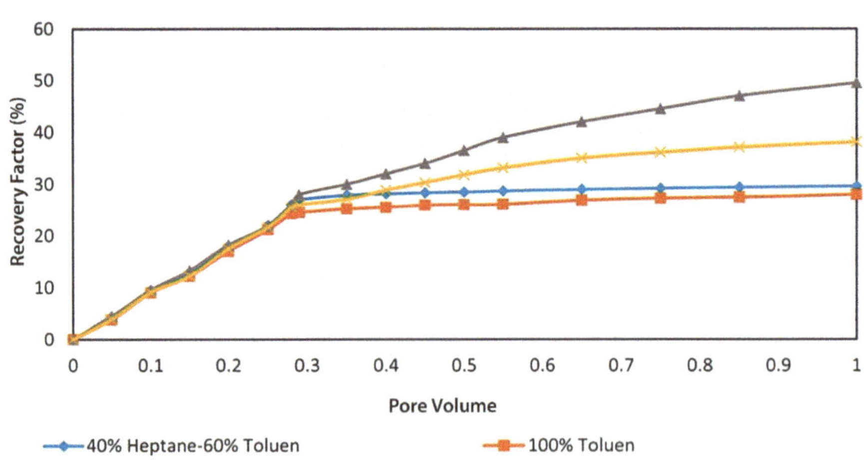

Fig. 5.24 Oil recovery factor vs. PV injected of 2000 ppm SiO_2 in brine to the micromodel saturated with two different fluids [139].

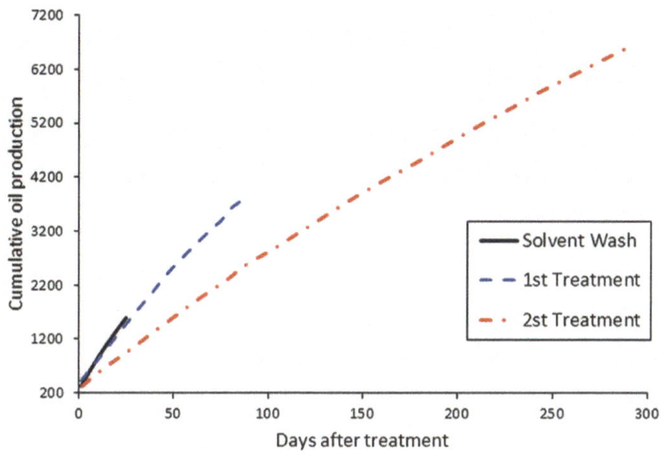

Fig. 5.25 The alteration of oil production after inhibitor treatment for the first and second time [104].

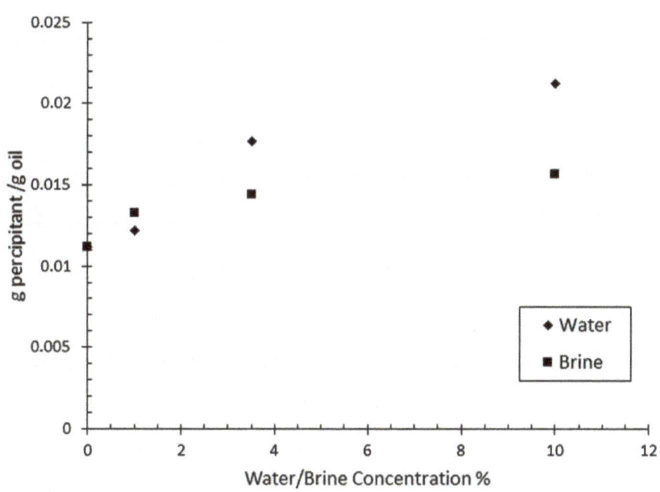

Fig. 5.26 Effect of the presence of water and brine with a different concentration on the influence of DBSA (50,000 ppm) as the asphaltene deposition inhibitor [94].

content decreases the adsorption of asphaltene on γ-Al_2O_3 nanoparticles. In addition, based on Fig. 5.26, the addition of water or brine causes greater deposition using an ionic surfactant; the influence of both water and brine decreases when their concentrations increases; however, there is no effect from water or brine using a nonionic surfactant [94].

5.2.3 Effects of inhibitors in different flow setups

In the dynamic inhibition tests, the effectiveness of the inhibitors was investigated in porous media (e.g., cores, packed beds, and microfluidics), and flow loops. In these experiments, the petrophysical properties such as permeability, porosity, relative permeability, wettability, and the treatment durability were tested directly or indirectly, using different inhibitors. In this section, we present the results of dynamic inhibition tests.

5.2.3.1 Core

Cores can be prepared from outcrops or hydrocarbon reservoir formations. Although the outcrop cores might have been eroded, they represent the reservoir rock when formation cores are not accessible. Formation cores are provided by coring operations during the drilling of target formations. Kashefi et al. [28] used a commercial polymeric inhibitor (SPI) during CO_2 injection in a sandstone core and observed that a 500 mg/L inhibitor concentration is efficient to avoid significant asphaltene precipitation and deposition for 7 PV injection. Without an inhibitor, the core permeability and effluent asphaltene content decreased by 70% and 51%, respectively. Consequently, the pressure drop in the core increased by 2.5 times compared to the case of using an inhibitor (Fig. 5.27) [28]. According to Fig. 5.27, E1 and E2 show the behavior of the system without and with the inhibitor, respectively. Any changes to the petrophysical properties are attributed to the asphaltene deposition in the core.

Karambeigi et al. [72] used IR95 as an asphaltene inhibitor and tested its performance in a carbonate core, for which the results are shown in Fig. 5.28. The pressure drop increased initially, which was due to the core resistance against the live oil entrance and viscosity changes due to replacing the live oil with dead oil. After 3 PV injection, the pressure drop for the case with inhibitor was 170 psi; without an inhibitor, the pressure drop across the core increased to 270 psi (Fig. 5.28(left)). Since the initial permeability of the core was low (0.293 mD), its changes were not significant and even the contrast between the case with and without inhibitor was hard to identify. Nevertheless, Fig. 5.28(right) shows the permeability improvement in case of inhibitor usage [72].

Hashemi et al. [71] investigated the effectiveness of NiO nanoparticles, using a similar system, as studied by Karambeigi et al. [72]. The co-injection of the nanoparticle with miscible CO_2 reduced the rate of permeability reduction, which was related to a decrease in asphaltene deposition, from 20% to 2%; the oil recovery increased by 7% compared to the case without

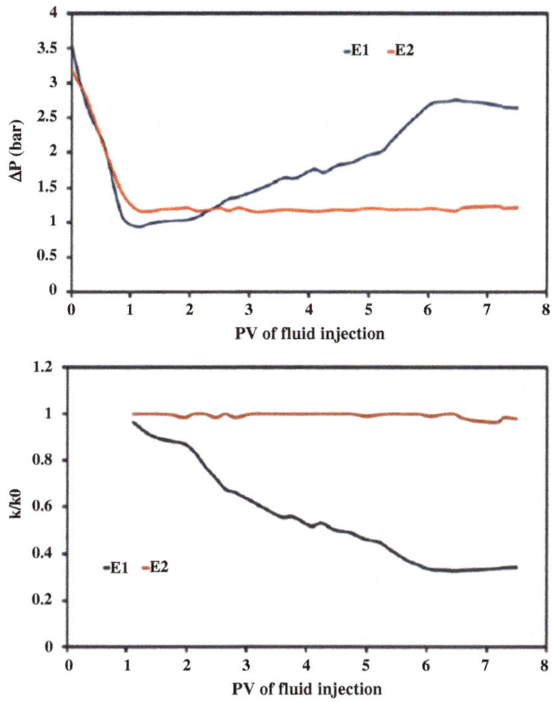

Fig. 5.27 An inhibitor effect during coreflood E1 plot shows flooding without inhibitor, and E2 plot depicts flooding with inhibitor. (left) Pressure drop and (right) permeability reduction [28].

an inhibitor [71]. Romero et al. investigated the effectiveness of the alumina nanoparticles in formation cores. They concluded that the nanoparticles reduce the asphaltene damage from 99% to less than 34%, and increase the permeability from 0.09 to 11 mD [17].

Madhi et al. [114] investigated the effectiveness of cetyl trimethyl ammonium bromide (CTAB) and sodium dodecyl sulfate (SDS) as the inhibitors in a limestone core with two different crude oils. Both inhibitors reduced the rate of permeability reduction and increased the relative permeability of the tested oils even to a higher value than its original value before the asphaltene damage [114]. Zwitterionic liquid surfactant changed the relative permeability, wettability, and subsequently the recovery of light crude in formation cores with different permeabilities. The relative permeability of Bedford lime and formation rock increased by 10.2% and 4.9%, respectively [4]. Organic phase inhibitors, such as deasphalted oils and vegetable oils, are also used for asphaltene inhibition in cores. For instance, Monjezi et al. [73] showed that injection

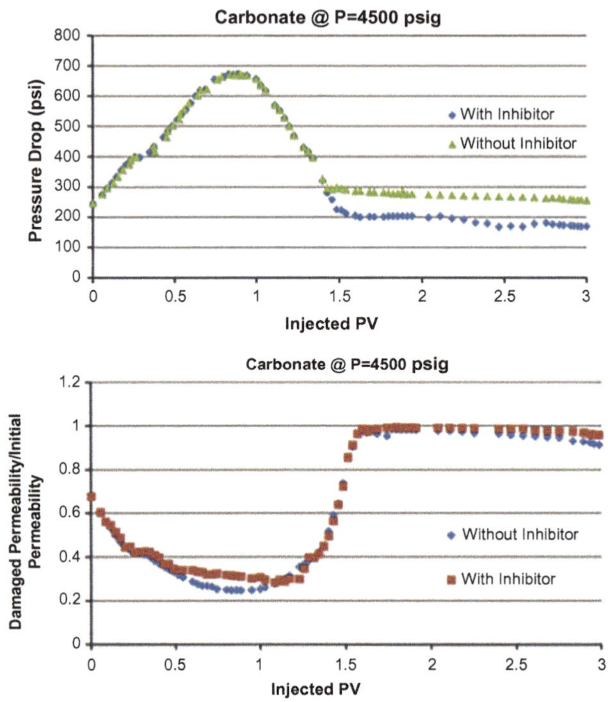

Fig. 5.28 (left) Pressure drop vs. PV of live oil, (right) permeability reduction factor vs. PV of live oil [72].

of a henna extract into a sandstone outcrop decreases the amount of asphaltene deposition and permeability reduction by 22.5% and 20%, respectively; the pressure drop across the core also decreases from 25 to 1 psi (Fig. 5.29).

5.2.3.2 Packed columns

Packed columns are usually used when reservoir rocks are not available or when the core particles do not have enough mechanical integrity to resist shearing. Piro et al. [140] studied two polymeric and two sulfonic inhibitors in asphaltene inhibition tests, using dolomite sand packs. All inhibitors reduced the asphaltene deposition, even after injecting 50 PV of the synthetic oil. The polymeric inhibitors were more effective compared to the sulfonic ones [140]. Betancur et al. [69] tested silica gel nanoparticles to assess their inhibitory effects in a sand pack; the oil recovery was increased by 11% compared to the case without an inhibitor. The residual oil saturations for the blank (neither precipitant nor inhibitor), damaged (precipitant added), and treated (both precipitant and inhibitor added) systems were reported as 21.4%, 25.8%, and 16.5%, respectively. The silica gel nanoparticles

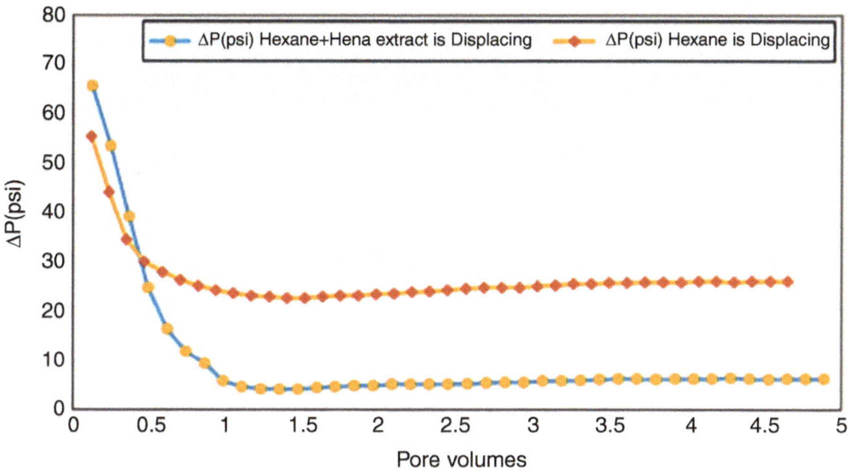

Fig. 5.29 Pressure drop in core as a result of nC_7 injection and *n*-heptane/henna extract injection [73].

reduced the residual oil saturation of the sand pack by increasing the oil's relative permeability and also through wettability alteration [69]. In another study, alumina nanofluid inhibited the asphaltene deposition in a silica sand pack. The oil effective permeability for the blank case was 521 mD, which was reduced to 390 mD for the packing, damaged by the asphaltene depositions. The nanofluid treatment increased the permeability up to 696 mD; the recovery was also increased by 7% compared to the damaged packing (Fig. 5.30) [70].

Kuang et al. [131] used three commercial inhibitors and three alkylphenol asphaltene inhibitors in a steel packed bed and studied the effect of the inhibitors on stabilizing the asphaltenes of different crude oils. Based on the type of crude oils, the inhibitors increased or decreased the asphaltene suspension [131]. Moreover, the material of the bead pack was also an important parameter in the efficiency of inhibitors. Al Sultan et al. [134] investigated the effect of glass and stainless steel surfaces on DBSA-asphaltene interaction. DBSA caused severe asphaltene deposition on stainless steel even more than the case without the inhibitor. For the glass bead case, the DBSA at low concentrations increased asphaltene deposition, and after a threshold limit, it reduced the amount of asphaltene deposition (Fig. 5.31). The rates of asphaltene deposition on stainless steel at 0, 10,000, and 25,000 ppm DBSA were 0.004, 0.0075, and 0.006 $g/cm^2/day$, respectively, and on glass, the rates of deposition at these concentration levels were 0.0065, 0.0045, and 0 $g/cm^2/day$ [134].

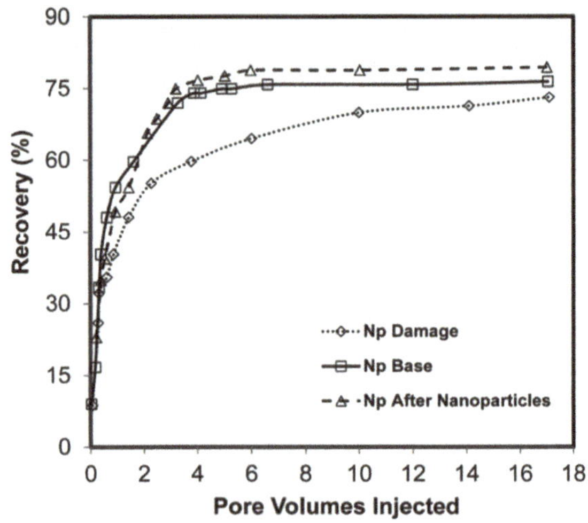

Fig. 5.30 Recovery curve for (1) base system, (2) NP injection simultaneously with crude followed by nC7 injection, (3) nC7 injection [70].

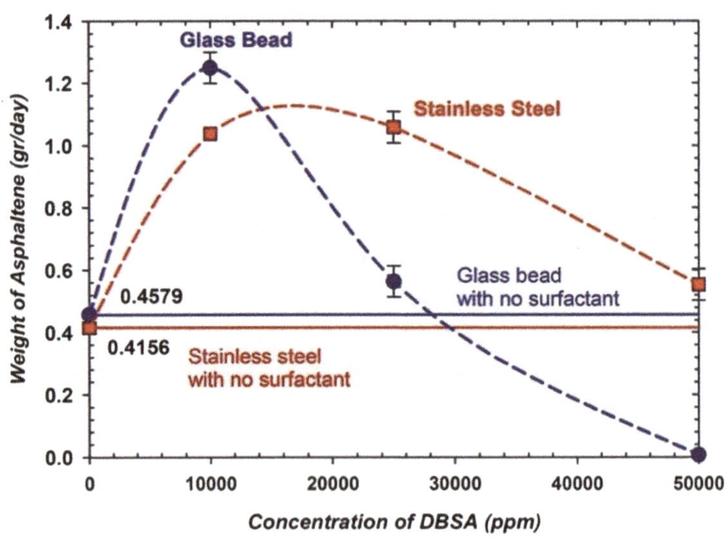

Fig. 5.31 Asphaltene deposited at 65 wt% nC7 flooded by a sample of asphaltene with 65 wt% but with different DBSA concentrations [134].

5.2.3.3 Microfluidics

A microfluidic device is a 2D (and sometimes 3D) network of pores that are usually made with glass or polydimethylsiloxane with a designated sketched pattern. Experiments in micromodels visualize the fluid flow and recovery mechanisms in a porous medium. With application to asphaltene inhibition

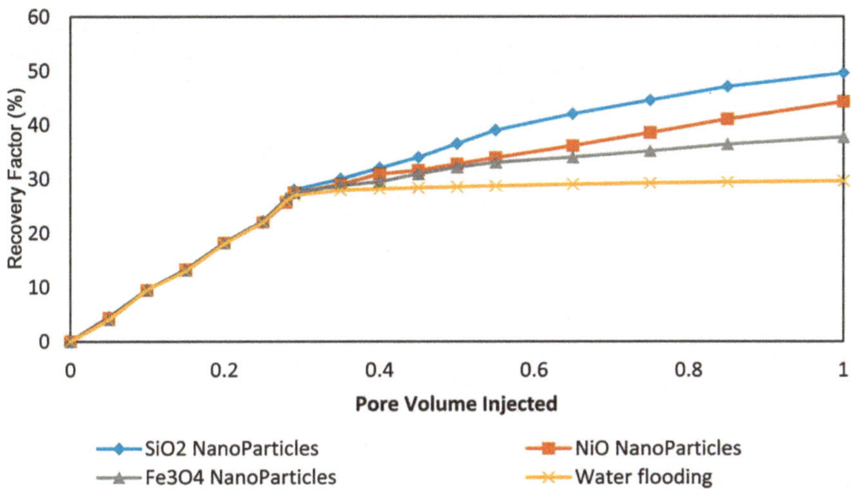

Fig. 5.32 Comparison of recovery factor due to the water flooding and three different nanofluids [139].

tests, Kazemzadeh et al. [139] investigated the effect of nanoparticles on oil recovery in a micromodel. The effective mechanism for the nanoparticles to inhibit asphaltene precipitation and deposition was wettability alteration of the glass microfluidic chip, IFT reduction, and mobility alteration between the displacing and displaced fluids. Fig. 5.32 shows that injection of 2000 ppm SiO_2, NiO, and Fe_3O_4 nanoparticles increases the recovery by 23%, 15%, and 8%, respectively (5000 ppm is the optimum for Fe_3O_4) [139].

Lin et al. [141] studied the effect of alkyl tails size in phenolic inhibitors on asphaltene deposition, using synthetic and crude oil. For synthetic oil, hexylphenol and octylphenol increased the amount and rate of asphaltene deposition, while dodecylphenol (DP) and isododecylphenol (isoDP) decreased the amount and rate of deposition compared to the baseline case. Surprisingly, in the case of crude injection, dodecylphenol and octylphenol caused more asphaltene deposition (which was even more severe for dodecylphenol) [141].

5.2.3.4 Flow loop and tubes

Hashmi and Firoozabadi [76] investigated the effectiveness of DBSA and toluene in the dissolution of deposited asphaltene in a stainless steel tube. The hydrocarbon flow rate did not affect the tube blockage by asphaltene depositions, while the type of inhibitor significantly affected the severity of the

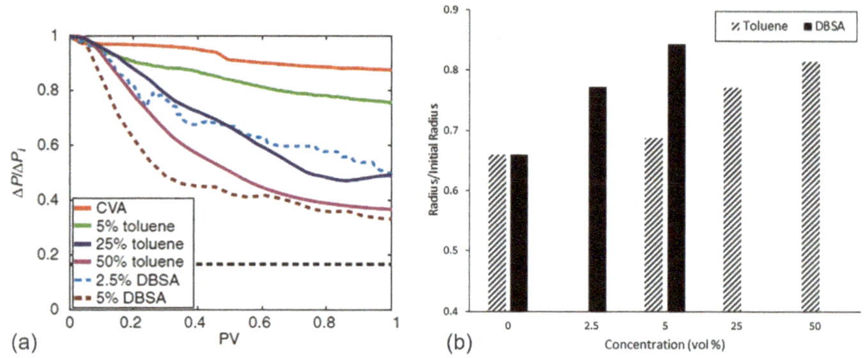

Fig. 5.33 Pressure drop reduction (A) and the radius of the capillary tube (B) after toluene and DBSA injection in different concentrations [76].

asphaltene deposition. Fig. 5.33 shows a better inhibitory performance of DBSA compared to toluene as it reduced pressure drop and asphaltene deposition by a greater amount, even at a lower concentration [76].

5.3 Final remarks

This chapter reviewed several experimental techniques to evaluate the effectiveness of inhibitors on asphaltene precipitation and deposition under different conditions, through static and dynamic tests. The asphaltene dispersant test is the simplest static method that can offer researchers a good perspective on the effectiveness of inhibitors. At the same time, NIR spectroscopy can show more specific results about asphaltene inhibitors. It should be noticed that different static methods measure inhibitor efficiency in different scales, which can cause controversy in making a conclusion. Each dynamic method has advantages, for example, the coreflood helps to study asphaltene inhibitor in a situation very similar to reservoirs while microfluidics approach helps to observe asphaltene-inhibitor interaction during the dynamic test. The effects of experiment variables such as temperature, pressure, type of operation, fluid properties, and rock properties on inhibition process were systematically discussed. Some inhibitors lost their efficiency with increasing temperature while the other's efficiency increased with increasing temperature. Fluid and asphaltene characteristics, including polarity and acidity, are important while ionic surfactant inhibitors are used. The dynamic tests showed that the asphaltene inhibitor efficiency also depends on rock properties, such as lithology, porosity, and permeability;

it can improve the reservoir rock properties, especially permeability, and fluid-rock properties, such as relative permeability and wettability. Therefore, both static and dynamic tests are necessary to evaluate inhibitors before pilot and field tests using target problematic oil and asphaltene, and the process conditions should be considered as similar as possible to the reservoir.

References

[1] R. Ortiz, et al., Asphaltene prevention workflow enhances oil production in HT fractured carbonate reservoirs, in: SPE International Conference and Exhibition on Formation Damage Control, Society of Petroleum Engineers, 2016.
[2] A. Albannay, T. Masuzawa, C.A. Dabbouk, Alternative methods in preventing asphaltene deposition, in: Abu Dhabi International Petroleum Exhibition and Conference, Society of Petroleum Engineers, 2010.
[3] H. Alboudwarej, D. Pole, W.Y. Svrcek, H.W. Yarranton, Adsorption of asphaltenes on metals, Ind. Eng. Chem. Res. 44 (15) (2005) 5585–5592.
[4] L.A. Alcázar-Vara, L.S. Zamudio-Rivera, E. Buenrostro-González, R.L. Hernández-Altamirano, V.Y. Mena-Cervantes, J.F. Ramírez-Pérez, Multifunctional properties of zwitterionic liquids. Application in enhanced oil recovery and asphaltene aggregation phenomena, Ind. Eng. Chem. Res. 54 (11) (2015) 2868–2878.
[5] T.A. Al-Sahhaf, M.A. Fahim, A.S. Elkilani, Retardation of asphaltene precipitation by addition of toluene, resins, deasphalted oil and surfactants, Fluid Phase Equilib. 194 (2002) 1045–1057.
[6] M. Boukherissa, F. Mutelet, A. Modarressi, A. Dicko, D. Dafri, M. Rogalski, Ionic liquids as dispersants of petroleum asphaltenes, Energy Fuel 23 (5) (2009) 2557–2564.
[7] C.-L. Chang, H.S. Fogler, Stabilization of asphaltenes in aliphatic solvents using alkylbenzene-derived amphiphiles. 1. Effect of the chemical structure of amphiphiles on asphaltene stabilization, Langmuir 10 (6) (1994) 1749–1757.
[8] C.-L. Chang, H. Scott Fogler, Peptization and coagulation of asphaltenes in apolar media using oil-soluble polymers, Fuel Sci. Technol. Int. 14 (1–2) (1996) 75–100.
[9] T.E. Chávez-Miyauchi, L.S. Zamudio-Rivera, V. Barba-López, E. Buenrostro-Gonzalez, J.M. Martínez-Magadán, N-aryl amino-alcohols as stabilizers of asphaltenes, Fuel 110 (2013) 302–309.
[10] T.E. Chávez-Miyauchi, L.S. Zamudio-Rivera, V. Barba-López, Aromatic polyisobutylene succinimides as viscosity reducers with asphaltene dispersion capability for heavy and extra-heavy crude oils, Energy Fuel 27 (4) (2013) 1994–2001.
[11] F.B. Cortés, J.M. Mejía, M.A. Ruiz, P. Benjumea, D.B. Riffel, Sorption of asphaltenes onto nanoparticles of nickel oxide supported on nanoparticulated silica gel, Energy Fuel 26 (3) (2012) 1725–1730.
[12] D. Dudášová, S. Simon, P.V. Hemmingsen, J. Sjöblom, Study of asphaltenes adsorption onto different minerals and clays: part 1. Experimental adsorption with UV depletion detection, Colloids Surf. A Physicochem. Eng. Asp. 317 (1–3) (2008) 1–9.
[13] A.M.A. Ghaffar, K.I. Kabel, R.K. Farag, N. Maysour, M.A. Zahran, Synthesis of poly (dodecyl phenol formaldehyde)-b-poly (oxypropylene) block copolymer, and evaluation as asphaltene inhibitor and dispersant, Res. Chem. Intermed. 41 (1) (2015) 443–455.
[14] S.M. Hashmi, A. Firoozabadi, Tuning size and electrostatics in non-polar colloidal asphaltene suspensions by polymeric adsorption, Soft Matter 7 (18) (2011) 8384–8391.
[15] N. Hosseinpour, A.A. Khodadadi, A. Bahramian, Y. Mortazavi, Asphaltene adsorption onto acidic/basic metal oxide nanoparticles toward in situ upgrading of reservoir oils by nanotechnology, Langmuir 29 (46) (2013) 14135–14146.

[16] P.M. McElfresh, D.L. Holcomb, D. Ector, Application of nanofluid technology to improve recovery in oil and gas wells, in: SPE International Oilfield Nanotechnology Conference and Exhibition, Society of Petroleum Engineers, 2012.
[17] Z. Romero, et al., Application and evaluation of a nanofluid containing nanoparticles for asphaltenes inhibition in well CPSXL4, in: OTC Brasil, Offshore Technology Conference, 2013.
[18] I.N. Evdokimov, N.Y. Eliseev, B.R. Akhmetov, Asphaltene dispersions in dilute oil solutions, Fuel 85 (10–11) (2006) 1465–1472.
[19] S.M. Hashmi, A. Firoozabadi, Effect of dispersant on asphaltene suspension dynamics: aggregation and sedimentation, J. Phys. Chem. B 114 (48) (2010) 15780–15788.
[20] S.M. Hashmi, A. Firoozabadi, Controlling nonpolar colloidal asphaltene aggregation by electrostatic repulsion, Energy Fuel 26 (7) (2012) 4438–4444.
[21] S.M. Hashmi, L.A. Quintiliano, A. Firoozabadi, Polymeric dispersants delay sedimentation in colloidal asphaltene suspensions, Langmuir 26 (11) (2010) 8021–8029.
[22] S.M. Hashmi, K.X. Zhong, A. Firoozabadi, Acid–base chemistry enables reversible colloid-to-solution transition of asphaltenes in non-polar systems, Soft Matter 8 (33) (2012) 8778–8785.
[23] M. Alhreez, D. Wen, L. Ali, A novel inhibitor for controlling Iraqi asphaltene problems, in: Environmental Impacts of the Oil and Gas Industries: Kurdistan Region of Iraq as a Case Study (EIOGI), 2017 International Conference on, IEEE, 2017, pp. 37–41.
[24] A. Aquino-Olivos, E. Buenrostro-Gonzalez, S.I. Andersen, C. Lira-Galeana, Investigations of inhibition of asphaltene precipitation at high pressure using bottomhole samples, Energy Fuel 15 (1) (2001) 236–240.
[25] C. Chen, J. Guo, N. An, Y. Pan, Y. Li, Q. Jiang, Study of asphaltene dispersion and removal for high-asphaltene oil wells, Pet. Sci. 9 (4) (2012) 551–557.
[26] H. Firoozinia, K.F.H. Abad, A. Varamesh, A comprehensive experimental evaluation of asphaltene dispersants for injection under reservoir conditions, Pet. Sci. 13 (2) (2016) 280–291.
[27] P. Juyal, V. Ho, A. Yen, S.J. Allenson, Reversibility of asphaltene flocculation with chemicals, Energy Fuel 26 (5) (2012) 2631–2640.
[28] S. Kashefi, A. Shahrabadi, M.N. Lotfollahi, A. Varamesh, A new polymeric additive as asphaltene deposition inhibitor in CO_2 core flooding, Korean J. Chem. Eng. 33 (11) (2016) 3273–3280.
[29] F. Marcano, L.G. Moura, F.M. Cardoso, P.T. Rosa, Evaluation of the chemical additive effect on asphaltene aggregation in dead oils: a comparative study between ultraviolet–visible and near-infrared-laser light scattering techniques, Energy Fuel 29 (5) (2015) 2813–2822.
[30] J. Marugán, J.A. Calles, J. Dufour, R. Giménez-Aguirre, J.L. Peña, D. Merino-García, Characterization of the asphaltene onset region by focused-beam laser reflectance: a tool for additives screening, Energy Fuel 23 (3) (2008) 1155–1161.
[31] O.C. Mullins, Asphaltenes in crude oil: absorbers and/or scatterers in the near-infrared region? Anal. Chem. 62 (5) (1990) 508–514.
[32] E. Rogel, O. Leon, Y. Espidel, Y. Gonzalez, Asphaltene stability in crude oils, SPE Prod. Facil. 16 (02) (2001) 84–88.
[33] I. Auflem, T. Havre, J. Sjöblom, Near-IR study on the dispersive effects of amphiphiles and naphthenic acids on asphaltenes in model heptane-toluene mixtures, Colloid Polym. Sci. 280 (8) (2002) 695–700.
[34] . https://aavos.eu/glossary/uv-vis-spectrometry/.
[35] C.-L. Chang, H.S. Fogler, Stabilization of asphaltenes in aliphatic solvents using alkylbenzene-derived amphiphiles. 2. Study of the asphaltene-amphiphile interactions and structures using Fourier transform infrared spectroscopy and small-angle X-ray scattering techniques, Langmuir 10 (6) (1994) 1758–1766.

[36] D. Weitz, J. Huang, M. Lin, J. Sung, Dynamics of diffusion-limited kinetic aggregation, Phys. Rev. Lett. 53 (17) (1984) 1657.
[37] M. Anisimov, et al., Asphaltene aggregation in hydrocarbon solutions studied by photon correlation spectroscopy, J. Phys. Chem. 99 (23) (1995) 9576–9580.
[38] Y.G. Burya, I.K. Yudin, V.A. Dechabo, V.I. Kosov, M.A. Anisimov, Light-scattering study of petroleum asphaltene aggregation, Appl. Optics 40 (24) (2001) 4028–4035.
[39] Y. Burya, I. Yudin, V. Dechabo, M. Anisimov, Colloidal properties of crude oils studied by dynamic light-scattering, Int. J. Thermophys. 22 (5) (2001) 1397–1410.
[40] B.J. Berne, R. Pecora, Dynamic Light Scattering with Application in Physics, Chemistry and Biology, John Wiley & Sons, New York, NY, USA, 1976.
[41] P. Clarke, B. Pruden, The development of an onset-of-precipitation detection technique using heat transfer analysis, Fuel Sci. Technol. Int. 14 (1–2) (1996) 117–137.
[42] P.F. Clarke, B.B. Pruden, Asphaltene precipitation: detection using heat transfer analysis, and inhibition using chemical additives, Fuel 76 (7) (1997) 607–614.
[43] M. del Carmen Garcia, H. Magaly, O. Jose, Asphaltene deposition prediction and control in a Venezuelan North Monagas oil field, in: International Symposium on Oilfield Chemistry, Society of Petroleum Engineers, 2003.
[44] M. Manek, Asphaltene dispersants as demulsification aids, in: SPE International Symposium on Oilfield Chemistry, Society of Petroleum Engineers, 1995.
[45] Y. Kazemzadeh, M. Malayeri, M. Riazi, R. Parsaei, Impact of Fe3O4 nanoparticles on asphaltene precipitation during CO_2 injection, J. Nat. Gas Sci. Eng. 22 (2015) 227–234.
[46] D.N. Rao, J. Lee, Application of the new vanishing interfacial tension technique to evaluate miscibility conditions for the Terra Nova offshore project, J. Petrol. Sci. Eng. 35 (3–4) (2002) 247–262.
[47] M. Nobakht, S. Moghadam, Y. Gu, Determination of CO_2 minimum miscibility pressure from measured and predicted equilibrium interfacial tensions, Ind. Eng. Chem. Res. 47 (22) (2008) 8918–8925.
[48] X. Wang, S. Zhang, Y. Gu, Four important onset pressures for mutual interactions between each of three crude oils and CO_2, J. Chem. Eng. Data 55 (10) (2010) 4390–4398.
[49] A. Zolghadr, M. Escrochi, S. Ayatollahi, Temperature and composition effect on CO2 miscibility by interfacial tension measurement, J. Chem. Eng. Data 58 (5) (2013) 1168–1175.
[50] T.W. Mattioli, J.R. Schieber, Deposition Test Coupon, Google Patents, 1979.
[51] J.S. Buckley, Microscopic investigation of the onset of asphaltene precipitation, Fuel Sci. Technol. Int. 14 (1–2) (1996) 55–74.
[52] A. Hirschberg, L. DeJong, B. Schipper, J. Meijer, Influence of temperature and pressure on asphaltene flocculation, Soc. Pet. Eng. J. 24 (03) (1984) 283–293.
[53] L.C.R. Junior, M.S. Ferreira, A.C. da Silva Ramos, Inhibition of asphaltene precipitation in Brazilian crude oils using new oil soluble amphiphiles, J. Petrol. Sci. Eng. 51 (1–2) (2006) 26–36.
[54] A. Rabbani, M. Gazanfari, I. Najafi, A new analytical modeling of asphaltene-induced formation damage under ultrasonic stimulation, in: 6th Congress of the Balkan Geophysical Society, 2011.
[55] A. Soleymanzadeh, M. Yousefi, S. Kord, O. Mohammadzadeh, A review on methods of determining onset of asphaltene precipitation, J. Pet. Explor. Prod. Technol. (2018) 1–22.
[56] J. Wang, J. Buckley, A two-component solubility model of the onset of asphaltene flocculation in crude oils, Energy Fuel 15 (5) (2001) 1004–1012.
[57] T. Maqbool, Understanding the Kinetics of Asphaltene Precipitation from Crude Oils (PhD thesis), The University of Michigan, 2011.

[58] M.J. Hÿtch, A.M. Minor, Observing and measuring strain in nanostructures and devices with transmission electron microscopy, MRS Bull. 39 (2) (2014) 138–146.
[59] O.C. Mullins, E.Y. Sheu, A. Hammami, A.G. Marshall, Asphaltenes, Heavy Oils, and Petroleomics, Springer Science & Business Media, 2007.
[60] L. Goual, M. Sedghi, X. Wang, Z. Zhu, Asphaltene aggregation and impact of alkylphenols, Langmuir 30 (19) (2014) 5394–5403.
[61] A. Einstein, Investigations on the Theory of the Brownian Movement, Courier Corporation, 1956.
[62] H. Goldsmith, S. Mason, The flow of suspensions through tubes. III. Collisions of small uniform spheres, Proc. R. Soc. Lond. A 282 (1391) (1964) 569–591.
[63] N. Graham, Orthokinetic flocculation rates for amorphous silica microspheres with cationic polyelectrolytes, Colloids Surf. 3 (1) (1981) 61–77.
[64] A. Hirschberg, Role of asphaltenes in compositional grading of a reservoir's fluid column, J. Petrol. Tech. 40 (01) (1988) 89–94.
[65] P. Sherman, The flow properties of emulsions, J. Pharm. Pharmacol. 16 (1) (1964) 1–25.
[66] J. Escobedo, G.A. Mansoori, Viscometric determination of the onset of asphaltene flocculation: a novel method, SPE Prod. Facil. 10 (02) (1995) 115–118.
[67] M. Shadman, A.S. Dehaghani, M.V. Sefti, M. Dehghanizadeh, The effect of inhibitors on asphaltene precipitation in crude oil using the viscometric method, Energy Sources, Part A 34 (9) (2012) 827–838.
[68] S.F. Alkafeef, F. Al-Medhadi, A.D. AL-Shammari, Asphaltene remedial technology using advanced deasphalted oil, in: Middle East Oil Show, Society of Petroleum Engineers, 2003.
[69] S. Betancur, J.C. Carmona, N.N. Nassar, C.A. Franco, F.B. Cortés, Role of particle size and surface acidity of silica gel nanoparticles in inhibition of formation damage by asphaltene in oil reservoirs, Ind. Eng. Chem. Res. 55 (21) (2016) 6122–6132.
[70] C.A. Franco, N.N. Nassar, M.A. Ruiz, P. Pereira-Almao, F.B. Cortés, Nanoparticles for inhibition of asphaltenes damage: adsorption study and displacement test on porous media, Energy Fuel 27 (6) (2013) 2899–2907.
[71] S.I. Hashemi, B. Fazelabdolabadi, S. Moradi, A.M. Rashidi, A. Shahrabadi, H. Bagherzadeh, On the application of NiO nanoparticles to mitigate in situ asphaltene deposition in carbonate porous matrix, Appl. Nanosci. 6 (1) (2016) 71–81.
[72] M.A. Karambeigi, M. Nikazar, R. Kharrat, Experimental evaluation of asphaltene inhibitors selection for standard and reservoir conditions, J. Petrol. Sci. Eng. 137 (2016) 74–86.
[73] E. Monjezi, A. Rostami, E. Joonaki, S.R. Shadizadeh, New application of henna extract as an asphaltene inhibitor: an experimental study, Asia Pac. J. Chem. Eng. 11 (6) (2016) 1027–1034.
[74] M. Madhi, A. Bemani, A. Daryasafar, M.R. Khosravi Nikou, Experimental and modeling studies of the effects of different nanoparticles on asphaltene adsorption, Pet. Sci. Technol. 35 (3) (2017) 242–248.
[75] A. Telmadarreie, Dynamic behavior of asphaltene precipitation and distribution pattern in carbonate reservoirs during solvent injection: pore-scale observations, in: SPE Canada Heavy Oil Technical Conference, Society of Petroleum Engineers, 2017.
[76] S.M. Hashmi, A. Firoozabadi, Effective removal of asphaltene deposition in metal-capillary tubes, SPE J. 21 (05) (2016) 1747–1754.
[77] S. Peramanu, P.F. Clarke, B.B. Pruden, Flow loop apparatus to study the effect of solvent, temperature and additives on asphaltene precipitation, J. Petrol. Sci. Eng. 23 (2) (1999) 133–143.
[78] N.N. Nassar, Asphaltene adsorption onto alumina nanoparticles: kinetics and thermodynamic studies, Energy Fuel 24 (8) (2010) 4116–4122.

[79] C. Ovalles, E. Rogel, H. Morazan, K. Chen, M.E. Moir, The use of nonylphenol formaldehyde resins for preventing asphaltene precipitation in vacuum residues and hydroprocessed petroleum samples, Pet. Sci. Technol. 34 (4) (2016) 379–385.
[80] C. Ovalles, E. Rogel, H. Morazan, M.E. Moir, Synthesis, characterization, and mechanism of asphaltene inhibition of phosphopropoxylated asphaltenes, Fuel 180 (2016) 20–26.
[81] E. Zahedi, M. Vafaie-Sefti, M. Shadman, H. Naderi, M. Amiri, A. Noorbakhsh, Experimental investigation of dodecylbenzene sulfonic acid and toluene dispersants on asphaltene precipitation of dead and live oil, Pet. Sci. Technol. 35 (7) (2017) 653–660.
[82] A.S. Ogunlaja, E. Hosten, Z.R. Tshentu, Dispersion of asphaltenes in petroleum with ionic liquids: evaluation of molecular interactions in the binary mixture, Ind. Eng. Chem. Res. 53 (48) (2014) 18390–18401.
[83] P. Permsukarome, C. Chang, H.S. Fogler, Kinetic study of asphaltene dissolution in amphiphile/alkane solutions, Ind. Eng. Chem. Res. 36 (9) (1997) 3960–3967.
[84] E.F. Ghloum, A.M. Rashed, M.A. Safa, R.C. Sablit, S.M. Al-Jouhar, Mitigation of asphaltenes precipitation phenomenon via chemical inhibitors, J. Petrol. Sci. Eng. 175 (2019) 495–507.
[85] D.W. Jennings, K.-P. Chao, J. Kim, Asphaltene inhibitor testing: comparison between a high pressure live-fluid deposition and ambient pressure dead-oil asphaltene stability method, in: Offshore Technology Conference, Offshore Technology Conference, 2018.
[86] K. Gharbi, K. Benyounes, C. Benamara, Inhibition of asphaltene flocculation in light crude oil: effect of pressure, temperature and inhibitor concentration, Pet. Sci. Technol. 36 (18) (2018) 1471–1477.
[87] A. Khormali, A.R. Sharifov, D.I. Torba, The control of asphaltene precipitation in oil wells, Pet. Sci. Technol. 36 (6) (2018) 443–449.
[88] T. Lu, Z. Li, W. Fan, X. Zhang, Q. Lv, Nanoparticles for inhibition of asphaltenes deposition during CO2 flooding, Ind. Eng. Chem. Res. 55 (23) (2016) 6723–6733.
[89] M. Igder, N. Hosseinpour, A. Amrollahi Biyouki, A. Bahramian, Control of asphaltenes aggregation in reservoir model oils along production streamline by Fe3O4 and NiO nanoparticles, Energy Fuel 32 (2018) 6689–6697.
[90] O. León, E. Rogel, A. Urbina, A. Andújar, A. Lucas, Study of the adsorption of alkyl benzene-derived amphiphiles on asphaltene particles, Langmuir 15 (22) (1999) 7653–7657.
[91] O. León, E. Contreras, E. Rogel, Amphiphile adsorption on asphaltene particles: adsorption isotherms and asphaltene stabilization, Colloids Surf. A Physicochem. Eng. Asp. 189 (1–3) (2001) 123–130.
[92] J. Wang, C. Li, L. Zhang, G. Que, Z. Li, The properties of asphaltenes and their interaction with amphiphiles, Energy Fuel 23 (7) (2009) 3625–3631.
[93] C.R. Mansur, A.R. de Melo, E.F. Lucas, Determination of asphaltene particle size: influence of flocculant, additive, and temperature, Energy Fuel 26 (8) (2012) 4988–4994.
[94] D. Subramanian, A. Firoozabadi, Effect of surfactants and water on inhibition of asphaltene precipitation and deposition, in: Abu Dhabi International Petroleum Exhibition and Conference, Society of Petroleum Engineers, 2015.
[95] S. Subramanian, S. Simon, J. Sjöblom, Interaction between asphaltenes and fatty-alkylamine inhibitor in bulk solution, J. Dispers. Sci. Technol. 39 (2) (2018) 163–173.
[96] L. Goual, A. Firoozabadi, Effect of resins and DBSA on asphaltene precipitation from petroleum fluids, AICHE J. 50 (2) (2004) 470–479.

[97] S. Kashefi, A. Shahrabadi, S. Jahangiri, M.N. Lotfollahi, H. Bagherzadeh, Investigation of the performance of several chemical additives on inhibition of asphaltene precipitation, Energy Sources, Part A 38 (24) (2016) 3647–3652.

[98] M. Mohammadi, M. Akbari, Z. Fakhroueian, A. Bahramian, R. Azin, S. Arya, Inhibition of asphaltene precipitation by TiO_2, SiO_2, and ZrO_2 nanofluids, Energy Fuel 25 (7) (2011) 3150–3156.

[99] O. Leon, E. Contreras, E. Rogel, G. Dambakli, J. Espidel, S. Acevedo, The influence of the adsorption of amphiphiles and resins in controlling asphaltene flocculation, Energy Fuel 15 (5) (2001) 1028–1032.

[100] K. Kraiwattanawong, H.S. Fogler, S.G. Gharfeh, P. Singh, W.H. Thomason, S. Chavadej, Effect of asphaltene dispersants on aggregate size distribution and growth, Energy Fuel 23 (3) (2009) 1575–1582.

[101] A.A. Melendez-Alvarez, et al., On the evaluation of the performance of asphaltene dispersants, Fuel 179 (2016) 210–220.

[102] A. Khormali, Asphaltene precipitation and inhibition in carbonate reservoirs, Pet. Sci. Technol. 35 (5) (2017) 515–521.

[103] S. Subramanian, L. Buscetti, S. Simon, M. Sacré, J. Sjöblom, Influence of fatty-alkylamine amphiphile on the asphaltene adsorption/deposition at the solid/liquid interface under precipitating conditions, Energy Fuel 32 (4) (2018) 4772–4782.

[104] S.J. Allenson, M.A. Walsh, A novel way to treat asphaltene deposition problems found in oil production, in: International Symposium on Oilfield Chemistry, Society of Petroleum Engineers, 1997.

[105] Y.R. Yin, A.T. Yen, S. Asomaning, Asphaltene inhibitor evaluation in CO_2 floods: laboratory study and field testing, in: SPE Permian Basin Oil and Gas Recovery Conference, Society of Petroleum Engineers, 2000.

[106] M. Means, Novel method for monitoring and controlling asphaltene deposition in crude oil, in: SPE Western Regional Meeting, Society of Petroleum Engineers, 2001.

[107] S. Al-Qabandi, A. Al-Naqi, Field trial of downhole asphaltene remediation treatment in a deep carbonate reservoir, Kuwait, in: Canadian International Petroleum Conference, Petroleum Society of Canada, 2003.

[108] A. Sanada, Y. Miyagawa, A case study of a successful chemical treatment to mitigate asphaltene precipitation and deposition in light crude oil field, in: SPE Asia Pacific Oil & Gas Conference and Exhibition, Society of Petroleum Engineers, 2006.

[109] Y. Villard, F. Fajardo, A. Milne, Enhanced oil recovery using innovative asphaltene inhibitors in East Venezuela, in: SPE International Conference and Exhibition on Formation Damage Control, Society of Petroleum Engineers, 2016.

[110] A.C. da Silva Ramos, L. Haraguchi, F.R. Notrispe, W. Loh, R.S. Mohamed, Interfacial and colloidal behavior of asphaltenes obtained from Brazilian crude oils, J. Petrol. Sci. Eng. 32 (2–4) (2001) 201–216.

[111] M. Barcenas, P. Orea, Molar-mass distributions of asphaltenes in the presence of inhibitors: experimental and computer calculations, Energy Fuel 25 (5) (2011) 2100–2108.

[112] M. Wang, J. Kaufman, X. Chen, C. Sungail, Development and evaluation of nonionic polymeric surfactants as asphaltene inhibitors, in: SPE International Symposium on Oilfield Chemistry, Society of Petroleum Engineers, 2015.

[113] I.A. Wiehe, T.G. Jermansen, Design of synthetic dispersants for asphaltenes, Pet. Sci. Technol. 21 (3–4) (2003) 527–536.

[114] M. Madhi, R. Kharrat, T. Hamoule, Screening of inhibitors for remediation of asphaltene deposits: experimental and modeling study, Petroleum 4 (2017) 168–177.

[115] N. Carnahan, J.L. Salager, R. Anton, Effect of resins on stability of asphaltenes, in: Offshore Technology Conference, Offshore Technology Conference, 2007.

[116] E.F. Ghloum, M. Al-Qahtani, A. Al-Rashid, Effect of inhibitors on asphaltene precipitation for Marrat Kuwaiti reservoirs, J. Petrol. Sci. Eng. 70 (1) (2010) 99–106.

[117] R.S. Mohamed, W. Loh, A.C. Ramos, C.C. Delgado, V.R. Almeida, Reversibility and inhibition of asphaltene precipitation in Brazilian crude oils, Pet. Sci. Technol. 17 (7–8) (1999) 877–896.
[118] N.N. Nassar, A. Hassan, P. Pereira-Almao, Metal oxide nanoparticles for asphaltene adsorption and oxidation, Energy Fuel 25 (3) (2011) 1017–1023.
[119] P.M. Spiecker, K.L. Gawrys, C.B. Trail, P.K. Kilpatrick, Effects of petroleum resins on asphaltene aggregation and water-in-oil emulsion formation, Colloids Surf. A Physicochem. Eng. Asp. 220 (1–3) (2003) 9–27.
[120] D. Merino-Garcia, S.I. Andersen, Interaction of asphaltenes with nonylphenol by microcalorimetry, Langmuir 20 (4) (2004) 1473–1480.
[121] L.Z. Pillon, Effect of dispersants and flocculants on the colloidal stability of asphaltenes, Pet. Sci. Technol. 19 (7–8) (2001) 863–873.
[122] A.F. Lima, C.R. Mansur, E.F. Lucas, G. González, Polycardanol or sulfonated polystyrene as flocculants for asphaltene dispersions, Energy Fuel 24 (4) (2009) 2369–2375.
[123] L.F.B. Moreira, E.F. Lucas, G. González, Stabilization of asphaltenes by phenolic compounds extracted from cashew-nut shell liquid, J. Appl. Polym. Sci. 73 (1) (1999) 29–34.
[124] D.F. Smith, G.C. Klein, A.T. Yen, M.P. Squicciarini, R.P. Rodgers, A.G. Marshall, Crude oil polar chemical composition derived from FT− ICR mass spectrometry accounts for asphaltene inhibitor specificity, Energy Fuel 22 (5) (2008) 3112–3117.
[125] H.H. Ibrahim, R.O. Idem, Correlations of characteristics of Saskatchewan crude oils/asphaltenes with their asphaltenes precipitation behavior and inhibition mechanisms: differences between CO_2-and n-heptane-induced asphaltene precipitation, Energy Fuel 18 (5) (2004) 1354–1369.
[126] H.H. Ibrahim, R.O. Idem, CO_2-miscible flooding for three Saskatchewan crude oils: interrelationships between asphaltene precipitation inhibitor effectiveness, asphaltenes characteristics, and precipitation behavior, Energy Fuel 18 (3) (2004) 743–754.
[127] H.H. Ibrahim, R.O. Idem, Interrelationships between asphaltene precipitation inhibitor effectiveness, asphaltenes characteristics, and precipitation behavior during n-heptane (light paraffin hydrocarbon)-induced asphaltene precipitation, Energy Fuel 18 (4) (2004) 1038–1048.
[128] M. del Carmen García, N. Chiaravallo, Asphaltenes deposition control in Lake Maracaibo crude oil production, in: SPE International Symposium on Oilfield Chemistry, Society of Petroleum Engineers, 2001.
[129] A.M. Atta, A.O. Ezzat, M.M. Abdullah, A.I. Hashem, Effect of different families of hydrophobic anions of imadazolium ionic liquids on asphaltene dispersants in heavy crude oil, Energy Fuel 31 (8) (2017) 8045–8053.
[130] A.H.S. Dehaghani, M.H. Badizad, Inhibiting asphaltene precipitation from Iranian crude oil using various dispersants: experimental investigation through viscometry and thermodynamic modelling, Fluid Phase Equilib. 442 (2017) 104–118.
[131] J. Kuang, et al., Assessment of the performance of asphaltene inhibitors using a multisection packed bed column, Fuel 241 (2019) 247–254.
[132] S. Schantz, W. Stephenson, Asphaltene deposition: Development and application of polymeric asphaltene dispersants, in: SPE Annual Technical Conference and Exhibition, Society of Petroleum Engineers, 1991.
[133] L.M. Cenegy, Survey of successful world-wide asphaltene inhibitor treatments in oil production fields, in: SPE Annual Technical Conference and Exhibition, Society of Petroleum Engineers, 2001.
[134] A. Al Sultan, M. Zirrahi, H. Hassanzadeh, J. Abedi, Effect of the surfactant on asphaltene deposition on stainless-steel and glass surfaces, Energy Fuel 32 (4) (2018) 5635–5642.
[135] C.-L. Chang, H. Fogler, Asphaltene stabilization in alkyl solvents using oil-soluble amphiphiles, in: SPE International Symposium on Oilfield Chemistry, Society of Petroleum Engineers, 1993.

[136] J. Pereira, J. Delgado-Linares, A. Briones, M. Guevara, C. Scorzza, J.-L. Salager, The effect of solvent nature and dispersant performance on asphaltene precipitation from diluted solutions of instable crude oil, Pet. Sci. Technol. 29 (23) (2011) 2432–2440.
[137] M. Barcenas, P. Orea, E. Buenrostro-González, L.S. Zamudio-Rivera, Y. Duda, Study of medium effect on asphaltene agglomeration inhibitor efficiency, Energy Fuel 22 (3) (2008) 1917–1922.
[138] T.S. Loureiro, L.C.M. Palermo, L.S. Spinelli, Influence of precipitation conditions (n-heptane or carbon dioxide gas) on the performance of asphaltene stabilizers, J. Petrol. Sci. Eng. 127 (2015) 109–114.
[139] Y. Kazemzadeh, S.E. Eshraghi, K. Kazemi, S. Sourani, M. Mehrabi, Y. Ahmadi, Behavior of asphaltene adsorption onto the metal oxide nanoparticle surface and its effect on heavy oil recovery, Ind. Eng. Chem. Res. 54 (1) (2015) 233–239.
[140] G. Piro, L.B. Canonico, G. Galbariggi, L. Bertero, C. Carniani, Asphaltene adsorption onto formation rock: an approach to asphaltene formation damage prevention, SPE Prod. Facil. 11 (03) (1996) 156–160.
[141] Y.-J. Lin, et al., Characterizing asphaltene deposition in the presence of chemical dispersants in porous media micromodels, Energy Fuel 31 (11) (2017) 11660–11668.
[142] G. Fritschy, E. Papirer, Interactions between a bitumen, its components and model fillers, Fuel 57 (11) (1978) 701–704.
[143] G. Gonzalez, A. Middea, Peptization of asphaltene by various oil soluble amphiphiles, Colloids Surf. 52 (1991) 207–217.
[144] H. Lian, J.-R. Lin, T.F. Yen, Peptization studies of asphaltene and solubility parameter spectra, Fuel 73 (3) (1994) 423–428.
[145] D. Dudášová, A. Silset, J. Sjöblom, Quartz crystal microbalance monitoring of asphaltene adsorption/deposition, J. Dispers. Sci. Technol. 29 (1) (2008) 139–146.
[146] N.N. Nassar, M.E. Al-Jabari, M.M. Husein, Removal of asphaltenes from heavy oil by nickel nano and micro particle adsorbents, in: Proceedings of the IASTED International Conference, vol. 615, 2008, p. 808.
[147] N.N. Nassar, A. Hassan, P. Pereira-Almao, Effect of surface acidity and basicity of aluminas on asphaltene adsorption and oxidation, J. Colloid Interface Sci. 360 (1) (2011) 233–238.
[148] N.N. Nassar, A. Hassan, P. Pereira-Almao, Effect of the particle size on asphaltene adsorption and catalytic oxidation onto alumina particles, Energy Fuel 25 (9) (2011) 3961–3965.
[149] K. Safaie, A.R.S. Nazar, Evaluation of asphaltene inhibitors effect on aggregation coupled sedimentation process, J. Dispers. Sci. Technol. 35 (3) (2014) 329–337.
[150] F. Shojaati, M. Riazi, S.H. Mousavi, Z. Derikvand, Experimental investigation of the inhibitory behavior of metal oxides nanoparticles on asphaltene precipitation, Colloids Surf. A Physicochem. Eng. Asp. 531 (2017) 99–110.
[151] M. Bouts, R. Wiersma, H. Muijs, A. Samuel, An evaluation of new aspaltene inhibitors; laboratory study and field testing, J. Petrol. Tech. 47 (09) (1995) 782–787.
[152] P.F. CLARKE, B.B. PRUDEN, Asphaltene precipitation from cold Lake and athabasca bitumens, Pet. Sci. Technol. 16 (3–4) (1998) 287–305.
[153] E. Rogel, E. Contreras, O. Leon, An experimental theoretical approach to the activity of amphiphiles as asphaltene stabilizers, Pet. Sci. Technol. 20 (7–8) (2002) 725–739.
[154] Y.-F. Hu, T.-M. Guo, Effect of the structures of ionic liquids and alkylbenzene-derived amphiphiles on the inhibition of asphaltene precipitation from CO_2-injected reservoir oils, Langmuir 21 (18) (2005) 8168–8174.
[155] E. Joonaki, R. Burgass, B. Tohidi, Experimental and modelling study on application of new class of asphaltene inhibitors for enhanced oil recovery purposes: adsorption and wettability alteration, in: Abu Dhabi International Petroleum Exhibition & Conference, Society of Petroleum Engineers, 2016.

[156] H. Yonebayashi, X. Zhang, Emerging numerical modelling technique to evaluate asphaltene inhibitor efficiency during entire field life, in: SPE Asia Pacific Oil & Gas Conference and Exhibition, Society of Petroleum Engineers, 2016.
[157] H. Yonebayashi, Case study of asphaltene flow assurance risk mitigation-emerging numerical model to evaluate inhibitor efficiency, in: 78th EAGE Conference and Exhibition 2016, 2016.
[158] M. Mohammadi, M. Dadvar, B. Dabir, TiO_2/SiO_2 nanofluids as novel inhibitors for the stability of asphaltene particles in crude oil: mechanistic understanding, screening, modeling, and optimization, J. Mol. Liq. 238 (2017) 326–340.
[159] M.M. Shadman, M. Dehghanizadeh, A.H. Saeedi Dehaghani, M. Vafaie Sefti, N. Mokhtarian, An investigation of the effect of aromatic, anionic and nonionic inhibitors on the onset of asphaltene precipitation, J. Oil Gas Petrochem. Technol. 1 (2014) 17–28.
[160] D. Subramanian, K. Wu, A. Firoozabadi, Ionic liquids as viscosity modifiers for heavy and extra-heavy crude oils, Fuel 143 (2015) 519–526.
[161] M. Tavakkoli, et al., Indirect method: a novel technique for experimental determination of asphaltene precipitation, Energy Fuel 29 (5) (2015) 2890–2900.

CHAPTER SIX

Modeling and simulation investigations of asphaltene deposition control by chemical inhibitors

6.1 Thermodynamic models

There are two main theories (within the thermodynamic framework) that define the state of asphaltenes in solution: (1) solubility theory and (2) colloidal theory. In the solubility theory, the asphaltenes are considered as a dissolved component (liquid) in the crudes; asphaltene precipitation is identified through liquid-liquid or solid-liquid equilibria [1]. In the colloidal theory, the asphaltenes are considered solid colloidal particles and are presented along with other molecules in the fluids in the form of micelles. Micelles contain an asphaltene core, and a shell that is comprised of other molecules such as resins. In this theory, the asphaltene cores are stabilized in the fluid by shells, which can be resins or chemical additives [2].

There are three main categories of the thermodynamic models, including (1) fundamental, (2) equations of state (EoSs), and (3) empirical methods. The fundamental methods are based on the minimization of the Gibbs energy, where the molecules are known to be more stable at the lowest level of energy. In physical chemistry, when the molecules in two adjoining phases are at equilibrium (minimum Gibbs free energy), the equilibrium condition also applies for the chemical potential of the solution, which is the basis of developing different equations of state. The empirical methods are equations developed by fitting the equation to the experimental results, and sometimes they are not developed based on any specific theory. Methods are described more in the following sections of this chapter.

Asphaltene Deposition Control by Chemical Inhibitors
https://doi.org/10.1016/B978-0-323-90510-7.00001-X

6.1.1 Fundamental methods

Researchers use the colloidal theory and the Gibbs free energy (or its modifications) to develop models that predict the behavior of thermodynamic and kinetic inhibitors. The thermodynamic micellization model [3, 4] is a thermodynamic method in which the asphaltene molecules are considered as a core and are encapsulated by other molecules in the shell. When the concentration of the resin or amphiphile increases in the shell, the interfacial tension between the asphaltene core and shell reduces, which stabilizes the asphaltene. Based on this model, the Gibbs free energy of the system is the summation of the contributions from the liquid and solid Gibbs free energy. The liquid phase Gibbs free energy is obtained by the summation of solvent and solute Gibbs free energies (Eq. 6.1). The solvent Gibbs free energy is obtained by the summation of the chemical potential of solvent molecules except for asphaltenes and resins. Eq. (6.2) shows the Gibbs free energy for solvents.

$$G^l = G_{solvent} + G_{solutes} \tag{6.1}$$

$$G_{solvent} = \sum_{i=1}^{Nc-4} N_i \mu_{ci} \tag{6.2}$$

where N_c is the number of components in the system. N_c, $N_c - 1$ and $N_c - 2$ represent a1 asphaltene C_{10}^{+}, a2 asphaltene C_7–C_{10} and a3 asphaltene C_5–C_7, respectively, and $N_c - 3$ refers to the resin or amphiphile. μ_c stands for the chemical potential of the component. The solute Gibbs free energy (Eq. 6.3) includes the Gibbs free energies at the standard state (G^0), mixing (G_M), and interactions (G_{int}) between the components, respectively.

$$G_{solutes} = G^0 + G_M + G_{int} \tag{6.3}$$

After the expansion of liquid Gibbs free energy and combining that with the Gibbs free energy of the solids (G^s), which is defined in Eq. (6.4), the Gibbs free energy of the system is calculated using Eq. (6.5).

$$G^s = \sum_{i=1}^{3} N_{ai}^s \left[\mu_{C_{ai}}^0(T) + kT\, ln x_{ai}^s f_{ai}^s\right] + N_r^s \left[\mu_{C_r}^0(T) + kT \ln x_r^s f_r^s\right] \tag{6.4}$$

$$G = \sum_{i=1}^{Nc} N_i \mu_{ci}^0(T) + kT \sum_{i=1}^{Nc-4} N_i \ln \hat{f}_i + kT \left(\sum_{i=1}^{3} N_{1ai} \ln x_{1ai} + N_{1r} \ln x_{1r} + N_m \ln x_m \right)$$

$$+ N_m \Delta G_m^{00} + kT \left(\sum_{i=1}^{3} N_{ai}^l \ln \phi_{ai} p + N_r^l \ln \phi_r p \right) + kT \left(\sum_{i=1}^{3} N_{ai}^s \ln x_{ai}^s f_{ai}^s + N_r^s \ln x_r^s f_r^s \right) \tag{6.5}$$

where μ^0 is the chemical potential of pure component; and x, f, and $\hat{f}$ refer to the mole fraction, fugacity of pure component, and fugacity of solvent component, respectively. k is the Boltzmann constant and T introduces the temperature. Subscripts of a, r, and m stand for the asphaltene, resin, and micelle, respectively. Superscript s and l refer to the solid phase and liquid phase, respectively. p denotes the pressure, ϕ symbolizes the fugacity coefficient, and $\Delta G^{00}{}_m$ represents the standard Gibbs free energy of micellization of a micelle which is defined by the following equation:

$$\Delta G_m^{00} = \left[(\Delta G_a^0)_{Ass} + (\Delta G_a^0)_{Tr} + (\Delta G_a^0)_{Def} \right] + \left[(\Delta G_r^0)_{Adp} + (\Delta G_r^0)_{Tr} + (\Delta G_r^0)_{Def} \right]$$

$$+ \left[\left(\Delta G_f^0\right)_{Adp} + \left(\Delta G_f^0\right)_{Tr} + \left(\Delta G_f^0\right)_{Def} \right] + (\Delta G^0)_{Inf} + (\Delta G^0)_{Ste} \tag{6.6}$$

Subscripts of *Ass*, *Tr*, *Def*, *Adp*, *Inf*, and *Ste* refer to the association energy between molecules in the micellar core, transfer energy of molecules from bulk to core, deformation energy of molecules in the shell, adsorption energy of molecules onto the shell, interfacial energy of asphaltene core and solvent in the shell, and the steric repulsion energy amphiphile, respectively. The subscript f introduces the amphiphile phase.

6.1.2 Equation of state (EoS)

Based on the solubility theory, the asphaltene is considered as a soluble component in the oil. The cubic equation of states (Cubic EoS) is developed based on the solubility theory, and the asphaltene-oil phase equilibrium is evaluated with liquid-liquid or solid-liquid equilibria. In the research studies on hydrocarbon systems, the fluids are commonly modeled with two cubic equations of state, namely Soave-Redlich-Kwong (SRK) and Peng-Robinson (PR). However, these models do not consider the polar and association intermolecular interactions corresponding to the presence of the heteroatoms and hydrogen bonds, which are the main mechanisms involved in the asphaltene aggregation and inhibition. Therefore, cubic-plus-association

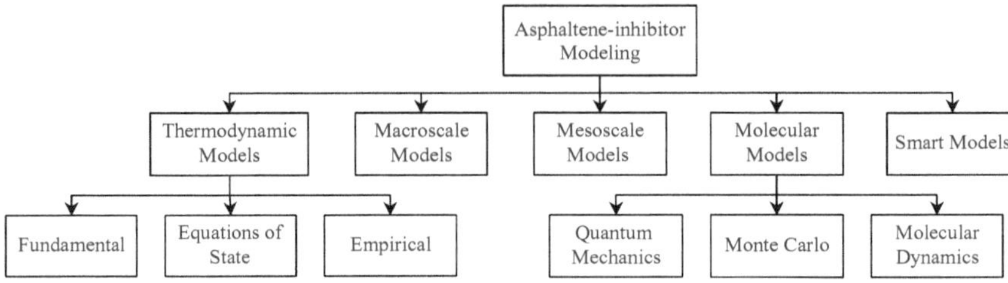

Fig. 6.1 Classification of different modeling approaches in asphaltene-inhibitor phase behavior.

(CPA) and perturbed chain association fluid theory (PC-SAFT) that both consider the association are developed and used for asphaltene-inhibitor modeling (Fig. 6.1). Kontogeorgis et al. introduced the CPA EoS using thermodynamics perturbation theory and considered the polar interactions such as self- and cross-association in the asphaltenes with themselves, or with the inhibitor molecules. The CPA model modifies the cubic EoSs such as the SRK and PR by adding contributions from association energy to those form of the physical hard-sphere interactions [5]. The CPA EoS is successful in predicting the onset point of the asphaltenes either in solid-liquid equilibria [6] or liquid-liquid equilibria [7]. Eqs. (6.7)–(6.23) show the mathematical framework of the CPA in case of considering either SRK or PR EoS for the physical hard-sphere interactions [5]. Eq. (6.7) introduces the compressibility factor of CPA, which is the summation of the compressibility factor of physical hard-sphere and association.

$$Z^{CPA} = Z^{Physical} + Z^{Association} \tag{6.7}$$

Eqs. (6.8)–(6.11) are used to calculate the compressibility factor and relative parameters based on the SRK EoS.

$$Z^{physical} = Z^{SRK} = \frac{\upsilon}{\upsilon - b} - \frac{a(T)}{RT(\upsilon + b)} \tag{6.8}$$

$$a(T) = 0.42747\alpha_{SRK}\frac{(RT_c)^2}{P_c} \tag{6.9}$$

$$\alpha_{SRK} = \left[1 + \left(0.48 + 1.574\omega - 0.176\omega^2\right)\left(1 - \sqrt{T_r}\right)\right]^2 \tag{6.10}$$

$$b_{SRK} = 0.08664\frac{RT_c}{P_c} \tag{6.11}$$

Eqs. (6.12)–(6.18) determine the compressibility factor and relative parameters using the PR EoS.

$$Z^{Physical} = Z^{PR} = \frac{v}{v-b} - \frac{a(T)v}{RT(v^2 + 2bv - b^2)} \tag{6.12}$$

$$a(T) = 0.45274\alpha_{PR}(T_r)\frac{(RT_c)}{P_c} \tag{6.13}$$

$$\alpha_{PR}(T_r) = \left[1 + m(1 - \sqrt{T_r}) - p(1 - T_r)(0.7 - T_r)\right]^2, \quad T_r \leq 1 \tag{6.14}$$

$$\alpha_{PR}(T_r) = \left[\exp\left(\left(1 - \frac{1}{d}\right)(1 - T_r^{\,d})\right)\right]^2, \quad T_r > 1 \tag{6.15}$$

$$b_{PR} = 0.07779\frac{RT_c}{P_c} \tag{6.16}$$

$$m = 0.37464 + 1.54226\omega - 0.26992\omega^2 \tag{6.17}$$

$$d = 1 + \frac{m}{2} + 0.3p \tag{6.18}$$

where v refers to the molar volume, b is the co-volume parameter, R represents the ideal gas constant, p denotes the pressure, and T introduces the temperature. The parameter $a(T)$ is an energy term, α is a parameter in the energy term, P_c symbolizes the critical pressure, and T_c indicates the critical temperature. ω stands for the acentric factor, and T_r is the temperature over critical temperature. Eqs. (6.19)–(6.23) introduce the compressibility factor and relative parameters that are used to calculate the association:

$$Z^{Association} = -\frac{1}{2}\left(1 + \frac{1}{v}\frac{\partial \ln g}{\partial\left(\frac{1}{v}\right)}\right)\sum_i x_i \sum_{A_i}(1 - X_{A_i}) \tag{6.19}$$

$$g(v) = \frac{1}{1 - 1.9\eta} \tag{6.20}$$

$$\eta = \left(\frac{b}{4v}\right) \tag{6.21}$$

$$X_{A_i} = \frac{1}{1 + \left(\frac{1}{v}\right)\sum_j x_j \sum_{B_j} X_{B_j}\Delta^{A_iB_j}} \tag{6.22}$$

$$\Delta^{A_iB_j} = g(v)\left[\exp\left(\frac{\varepsilon^{A_iB_j}}{RT}\right) - 1\right] b_{ij}\beta^{A_iB_j} \tag{6.23}$$

where $g(v)$ is a radial distribution function, x_i stands for the mole fraction of component i, X_{Ai} refers to the amount of asphaltene deposition calculated by the model, and η indicates the reduced density. Δ^{AB} denotes the association strength, ε^{AB} represents the association energy parameter, and β^{AB} denotes the association volume parameter between the sites A and B. For mixtures, a mixing rule is required to calculate a and b parameters and Mathias is one of the mixing rules, as given by the following equations:

$$a = \sum_i \sum_j x_i x_j \sqrt{a_i a_j}\left(1 - k_{ij}\right) \tag{6.24}$$

$$b = \sum_i x_i b_i \tag{6.25}$$

where k_{ij} is the binary interaction parameter. Gross and Sadowski [8] modified the statistical association fluid theory (SAFT) using perturbed chain (PC) and introduced PC-SAFT as another promising method for asphaltene modeling. Although there are extensive reports on asphaltene modeling using PC-SAFT, this method was not used widely to model the asphaltene-inhibitor behavior. In this method, the compressibility factor (Z) is the summation of the ideal (Z^{id}), hard chain (Z^{hc}), dispersion (Z^{disp}), and association (Z^{assoc}) compressibility factor (see Eq. 6.26) [9].

$$Z = Z^{id} + Z^{hc} + Z^{disp} + Z^{assoc} \tag{6.26}$$

The compressibility factor of the hard chain is defined by the following equation:

$$Z^{hc} = \overline{m} Z^{hs} - \sum_i x_i (m_i - 1) \dot{\rho} \left(g_{ij}^{hs}\right)^{-1} \frac{\partial g_{ij}^{hs}}{\dot{\rho}} \tag{6.27}$$

$$\overline{m} = \sum_i x_i m_i \tag{6.28}$$

where x_i and m_i refer to the mole fraction and segment number of component i. g_{ij}^{hs} stands for the average radial distribution function of the hard-sphere system (see Eq. 6.29); and Eq. (6.30) defines the expression $\dot{\rho}\frac{\partial g_{ij}^{hs}}{\partial \dot{\rho}}$.

$$g_{ij}^{hs} = \frac{1}{1-\zeta_3} + \frac{d_i d_j}{d_i + d_j} \frac{3\zeta_2}{(1-\zeta_3)^2} + \left(\frac{d_i d_j}{d_i + d_j}\right)^2 \frac{2\zeta_2^2}{(1-\zeta_3)^3} \tag{6.29}$$

$$\dot{\rho}\frac{\partial g_{ij}^{hs}}{\partial \dot{\rho}} = \frac{\zeta_3}{(1-\zeta_3)^2} + \frac{d_i d_j}{d_i + d_j}\left(\frac{3\zeta_2}{(1-\zeta_3)^2} + \frac{6\zeta_2\zeta_3}{(1-\zeta_3)^3}\right) + \left(\frac{d_i d_j}{d_i + d_j}\right)^2 \left(\frac{4\zeta_2^2}{(1-\zeta_3)^3} + \frac{6\zeta_2^2\zeta_3}{(1-\zeta_3)^4}\right) \tag{6.30}$$

where ζ_n refers to the packing factor and n can be [0,1,2,3]. d_i denotes the temperature-dependent segment diameter and $\dot{\rho}$ is the total number density of molecules (Eqs. 6.31–6.33).

$$\zeta_n = \frac{\pi}{6}\sum_i x_i m_i d_i^n \tag{6.31}$$

$$d_i = \sigma_i\left[1 - 0.12\exp\left(\frac{3\varepsilon_i}{kT}\right)\right] \tag{6.32}$$

$$\dot{\rho} = \frac{6\zeta_3}{\pi\sum_i x_i m_i d_i^3} \tag{6.33}$$

Eq. (6.34) defines the compressibility factor of the hard sphere.

$$Z^{hs} = \frac{\zeta_3}{1-\zeta_3} + \frac{3\zeta_1\zeta_3}{\zeta_0(1-\zeta_3)^2} + \frac{3\zeta_2^3 - 3\zeta_3\zeta_2^3}{\zeta_0(1-\zeta_3)^3} \tag{6.34}$$

Eq. (6.35) introduces the compressibility factor of displacement. Eqs. (6.36)–(6.39) demonstrate C_1 and C_2 coefficients, $\overline{m^2\varepsilon\sigma^3}$ and $\overline{m^2\varepsilon^2\sigma^3}$.

$$Z^{disp} = -2\pi\rho\frac{\partial(\zeta_3 I_1)}{\partial\zeta_3}\overline{m^2\varepsilon\sigma^3} - \pi\rho\overline{m}\left(C_1\frac{\partial(\zeta_3 I_2)}{\partial\zeta_3} + C_2\zeta_3 I_2\right)\overline{m^2\varepsilon^2\sigma^3} \tag{6.35}$$

$$C_1 = \left(1 + \overline{m}\frac{8\zeta_3 - 2\zeta_3^2}{(1-\zeta_3)^4} + (1-\overline{m})\frac{20\zeta_3 - 27\zeta_3^2 + 12\zeta_3^3 - 2\zeta_3^4}{((1-\zeta_3)(2-\zeta_3))^2}\right)^{-1} \tag{6.36}$$

$$C_2 = -C_1^2\left(\overline{m}\frac{-4\zeta_3^2 + 20\zeta_3 + 8}{(1-\zeta_3)^5} + (1-\overline{m})\frac{2\zeta_3^3 + 12\zeta_3^2 - 48\zeta_3 + 40}{((1-\zeta_3)(2-\zeta_3))^3}\right) \tag{6.37}$$

$$\overline{m^2\varepsilon\sigma^3} = \sum_{i=1}^{N}\sum_{j=1}^{N} x_i x_j m_i m_j\left(\frac{\varepsilon_{ij}}{kT}\right)\sigma_{ij}^3 \tag{6.38}$$

$$\overline{m^2\varepsilon^2\sigma^3} = \sum_{i=1}^{N}\sum_{j=1}^{N} x_i x_j m_i m_j\left(\frac{\varepsilon_{ij}}{kT}\right)^2\sigma_{ij}^3 \tag{6.39}$$

Eqs. (6.40)–(6.45) show the definition of relative parameters that are used in Eqs. (6.35), (6.38), and (6.39).

$$\varepsilon_{ij} = \sqrt{\varepsilon_i\varepsilon_j}\left(1 - k_{ij}\right) \tag{6.40}$$

$$\sigma_{ij} = \frac{1}{2}\left(\sigma_i + \sigma_j\right) \tag{6.41}$$

$$I_1 = \sum_{i=0}^{6} a_i(\overline{m})\zeta_3^i \tag{6.42}$$

$$I_2 = \sum_{i=0}^{6} b_i(\overline{m})\zeta_3^i \tag{6.43}$$

$$a_i(\overline{m}) = a_{0i} + \frac{\overline{m}-1}{\overline{m}}a_{1i} + \frac{\overline{m}-1}{\overline{m}}\frac{\overline{m}-2}{\overline{m}}a_{2i} \tag{6.44}$$

$$b_i(\overline{m}) = b_{0i} + \frac{\overline{m}-1}{\overline{m}}b_{1i} + \frac{\overline{m}-1}{\overline{m}}\frac{\overline{m}-2}{\overline{m}}b_{2i} \tag{6.45}$$

The compressibility factor of association is defined as follows:

$$Z^{assoc} = -\frac{1}{2}\left(1 + \rho\frac{\partial \ln g}{\partial \rho}\right)\sum_i x_i \sum_{A_i}\left(1 - X^{A_i}\right) \tag{6.46}$$

6.1.3 Empirical methods

Empirical models such as Langmuir, Freundlich, Temkin, and Dubinin-Radushkevich can be used to model the adsorption of molecules onto a surface, based on their underlying assumptions. The Langmuir model assumes a homogeneous surface and monolayer adsorption; the Freundlich model considers multilayer adsorption on a heterogeneous surface; the Temkin model assumes that the heat of adsorption for the molecules reduces linearly in each adsorbed layer; and the Dubinin-Radushkevich model assumes the adsorption of gases onto porous adsorbents through pore-filling mechanism. These adsorption models (isotherms) are used to predict the adsorption behavior of the asphaltenes (as an adsorbent particle) onto nanoparticles (as an adsorbate surface); this process is the primary mechanism for the nanoparticles to inhibit (or mitigate) the asphaltene deposition. Therefore, these empirical models are applicable to simulate the asphaltene-nanoparticle interaction. The formulations of the amount of adsorbed asphaltene onto nanoparticles (q_e [mg/g]) using different models are provided below:

Langmuir isotherm:

$$q_e = \frac{q_m K C_e}{1 + K C_e} \tag{6.47}$$

in which, K (L/mg) is the isotherm constants, C_e refers to the equilibrium concentration of the asphaltene (mg/L), and q_m introduces the maximum amount of asphaltene adsorbed onto nanoparticles (mg/g).

Freundlich isotherm:

$$q_e = K_f (C_e)^{1/n} \tag{6.48}$$

where K_f [(mg/g) (L/mg)$^{1/n}$] is the isotherm constants and $^1/_n$ indicates the adsorption intensity factor.

Temkin model:

$$q_e = B \ln A_T + B \ln C_e \tag{6.49}$$

$$B = \frac{RT}{b_T} \tag{6.50}$$

where A_T refers to the Temkin isotherm equilibrium binding constant (L/g), R is the ideal gas constant (8.314 J/mol K), T introduces the temperature (K), and b_T (J/mol) is isotherm constant.

Dubinin-Radushkevich isotherm:

$$q_e = q_s \exp\left(-K_{ad}\varepsilon^2\right) \tag{6.51}$$

$$\varepsilon = RT \ln\left[1 + \frac{1}{C_e}\right] \tag{6.52}$$

where q_s (mg/g) is the maximum adsorption capacity, K_{ad} (mol^2/kJ^2) is the isotherm constant, and ε (kJ/mol) denotes the adsorption potential.

6.2 Macroscale models

The effect of the asphaltene inhibitors on parameters such as permeability should be investigated under dynamic conditions to achieve a more realistic prediction. The dynamic tests have not received as much attention as the static tests. Darabi and Sepehrnoori [10] developed a comprehensive nonisothermal compositional reservoir simulator to address the reservoir performance upon asphaltene deposition and also asphaltene inhibition. Their 3D model used an implicit pressure and explicit composition (IMPEC) algorithm. They simulated the asphaltene deposition with inhibitor by considering adsorption, entrainment, and pore throat plugging; their model could incorporate the types of inhibitor, the concentration of inhibitor, soaking time, number of treatment cycles, and the turnover time period [10]. Fig. 6.2 shows a general flowchart for modeling of asphaltene deposition proposed by Miri et al. [11].

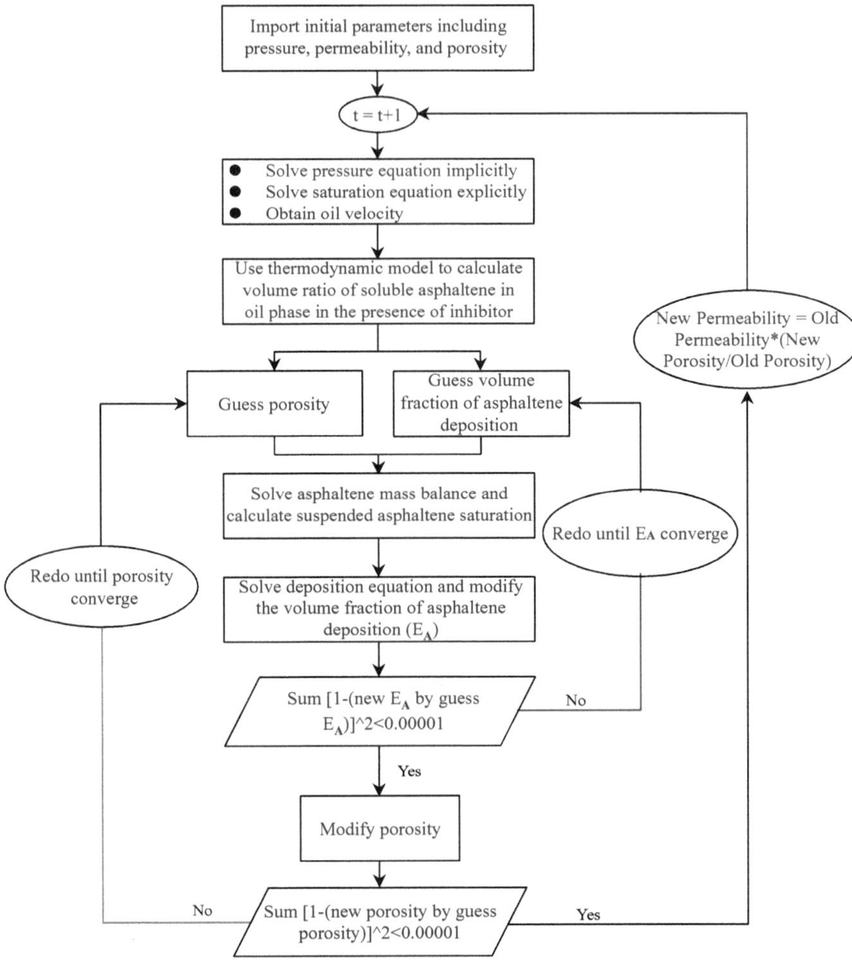

Fig. 6.2 Flowchart for the algorithm of simulating effect of inhibitor on asphaltene precipitation and desposition in porous media [11].

Empirical models are also introduced to correlate the permeability ratio (new permeability divided by the initial permeability) in porous media upon asphaltene deposition. The mechanisms of asphaltene deposition in porous media can be identified with or without the inhibitors using empirical models. The deposition mechanisms are gradual pore blocking, plugging of the pore throat, and/or in situ cake formation. Each mechanism changes the permeability ratio with a specific trend, as shown by Eqs. (6.53)–(6.55) [12].

Gradual pore blocking:

$$\sqrt{\frac{K}{K_i}} = 1 - a(PV) \tag{6.53}$$

Plugging of pore throat:

$$\frac{K}{K_i} = 1 - b(PV) \tag{6.54}$$

In situ cake formation:

$$\frac{K}{K_i} = 1 + c(PV) \tag{6.55}$$

where K refers to the permeability, K_i introduces the initial permeability, PV stands for the injected pore volume, and a, b, and c are the fitting parameters.

6.3 Mesoscale models

Multiphase flow in porous media can be modeled using macroscale computational fluid dynamic (CFD) methods through solving the Navier-Stokes equation. The validity of fluid flow simulation in porous media depends on constitutive relationships and semiempirical parameters such as Darcy law and relative permeability, respectively. Nevertheless, the CFD method struggles to model a moving boundary interface during phase change, fluid-solid interface effect such as wettability, and problems with complex geometries [13]. Lattice Boltzman method (LBM) was introduced in the 1980s as a developed version of the lattice gas cellular automation method to model multiphase and multicomponent flow in systems with complex geometries. LBM can also be coupled with EoSs such as CPA to predict the fluid-phase properties more accurately [14]. The LBM predicts the fluid dynamics behavior by considering the population movements of fictitious particles along a fixed Cartesian lattice [15]. The lattice Boltzman equation includes two steps: advection and collision (Eq. 6.56). The advection term represents the propagation of distribution function in the adjacent lattice along the velocity vector (left side of Eq. 6.56). The collision term represents the redistribution of particles across the velocity vector (right-hand side of Eq. 6.56) [13].

$$f(x + e_i\delta_t, t + \delta_t) - f(x, t) = \Omega(x, t) \tag{6.56}$$

where f is the single-particle distribution function, Ω is the collision operator, x is the position, and t refers to the time. The collision stage is the core of LBM algorithm as it preserves the flow physics. The Bhatnagar-Gross-Krook (BGK) collision operator is simple and effective, but it becomes unstable quickly for turbulent flow systems. Therefore, other operators, including multiple relaxation times are introduced [15].

During oil production, it is important to know the locus of asphaltene precipitation and deposition. The interactions between fluid and solid can be accounted, using LBM considering the bounce-back scheme for boundary conditions. For this matter, a few LBM variations have been introduced and utilized in the literature including interparticle potential model, free energy model, and stabilized diffuse-interface model. Fazeli et al. [16] developed a new probabilistic nucleation model to simulate mineral nucleation and growth in porous media. This mode was used for calcite nucleation on quartz and it can be alternatively used for asphaltene modeling. To the best of our knowledge, only Lin et al. [17] used LBM in asphaltene studies in which the deposited asphaltene was treated as a solid surface. They did not study the asphaltene precipitation and deposition occurring during the fluid flow. Therefore, there is no dynamic study on asphaltene aggregation, precipitation, and deposition using LBM neither with nor without the asphaltene inhibitor.

6.4 Molecular models

Molecular models estimate the properties and behaviors of a system by accounting for the interactions between different components at a molecular scale, for which the chemical structure of those components should be known. The molecular simulation studies compute the system potential energy from the position of the molecules. Other physical properties such as vapor pressure and viscosity can also be calculated by statistical thermodynamics. Molecular simulation can be used instead of the classical thermodynamic models (such as the EoS approach) for systems lacking experimental data. The molecular simulation includes quantum chemistry methods and statistical thermodynamics methods. In the quantum chemistry, the system can only access discrete energy levels while in the statistical thermodynamics, the system can possess any energy levels. The quantum chemistry model can be implemented with quantum mechanics simulation. The statistical thermodynamics can be implemented with Monte Carlo simulation and molecular dynamics. In the statistical thermodynamic simulation, an experimental condition can be modeled by fixing a set of temperature, pressure, volume, and number of molecules. Six statistical ensembles are defined, depending on the desired application (Table 6.1).

There are significant uncertainties about the chemical structures of the asphaltenes. According to the theory introduced by Mullins [19], there are two general types for the asphaltene chemical structures, namely island (or continental) and archipelago types; these structures were introduced in Chapter 1. So far, all the three mentioned molecular simulations are used to model the asphaltene-inhibitor interactions. In this chapter, we introduce the fundamentals of these three molecular models; then, we discuss the corresponding results from the literature studies.

6.4.1 Quantum mechanics

The quantum mechanics method is a mathematical approach based on quantum theory that models atoms with complex orbital shapes. The quantum mechanics technique is based on probability rather than certainty, and it is usually used to minimize the energy of a molecule and to optimize the geometry of structures. Density functional theory (DFT) is known as one of the effective algorithms in the framework of quantum mechanics that can be used to study the asphaltene-inhibitor interaction.

In an ideal system, the thermodynamic stability energy is evaluated by subtracting the mixture energy from the energy of individual components,

Table 6.1 Statistical ensembles used in molecular simulation [18].

Statistical ensemble	Imposed variables	Applications
Canonical ensemble	N, V, T	Phase properties (P, H, C_v, μ_i, ...)
Grand Canonical ensemble	μ_i, V, T	Adsorption isotherms, selectivity
Isothermal-isobaric ensemble	N, P, T	Phase properties (H, C_p, ρ, μ_i)
Microcanonical ensemble	E, V, T	Transport properties from molecular dynamics
Gibbs ensemble at imposed global volume (m phase)	$N = N_1 + N_2 + \ldots N_m$ $V = V_1 + V_2 + \ldots V_m$ T	Phase equilibrium of pure components and mixtures
Gibbs ensemble at imposed pressure (m phase)	$N = N_1 + N_2 + \ldots N_m$, P, T	Phase equilibrium of mixtures

which is called the binding energy (E_{bind}). This definition of the binding energy does not consider component deformation, which usually happens in two molecule interactions. Therefore, the interaction energy ($E_{\text{interaction}}$) is obtained by subtracting the restructuring energy of all components ($\sum \Delta E_{\text{fragment}}$) from the binding energy. The restructuring energy includes energy changes due to deformation and distortion. Eq. (6.57) shows the energy of a system for which the components have significant deformation during the interactions [20].

$$E_{\text{interaction}} = E_{\text{bind}} - \sum \Delta E_{\text{fragment}} \quad (6.57)$$

In the asphaltene-inhibitor system, the interaction energy (Eq. 6.57) is calculated with the dimeric or trimeric energy as the binding energy and the monomers energy as the fragment energy (Eq. 6.58) [21, 22].

$$\Delta E_{\text{interaction}} = \Delta E_{\text{dimer}} - \sum_{i=1}^{n} \Delta E_{\text{monomer,i}} \quad (6.58)$$

6.4.2 Monte Carlo

Monte Carlo simulation generates the configuration of systems that features the lowest energy based on the statistical methods and using a desired statistical ensemble. In this method, each molecule moves by a random chance to generate a new configuration. Based on the energy difference, ΔU, between the new and last accepted configuration, and the specific criterion for the energy difference, the new configuration is either accepted or rejected. The Metropolis algorithm is used for the acceptance and rejection criteria [23]. This algorithm can be biased toward generating configurations that are more probabilistically accepted, which is called the statistical bias method. In the Monte Carlo simulation, various types of moves are considered to try all possible molecular orientations, locations, internal configurations, and basically all configurations as described by Ungerer and Lachet with details [18]. In the Monte Carlo simulation, molecules are considered as rigid particles or surfaces for which molecular interactions (aggregation or suspension) are identified based on the distance between the particles. In the literature, usually, the particle agglomeration control (PAC) method is coupled with the Monte Carlo simulation to account for the asphaltene-asphaltene and the asphaltene-inhibitor interactions [24].

The Monte Carlo method uses time-driven, event-driven, and multidriven methods. Liffman [25] developed the time-driven model in 1992 in which all probable events occur at a fixed time step. This technique is suitable to be applied for the colloidal aggregation, which leads to the dynamic variations in the asphaltene aggregate size [26]. In 1987, Garcia et al. introduced the event-driven model in which an event occurs initially; then, time advances within the related time step [27]. The multi-driven method developed by Zhao et al. uses statistical weighting, which simulates the real particulate system accurately. This method is beneficial in the modeling of dispersed systems with coagulation and condensation [28].

6.4.3 Molecular dynamics

In the molecular dynamics simulation, the equation of motion is solved for every atom, and the position of each atom is obtained at every time step. In this modeling approach, the intramolecular potential is obtained by considering the contributions to the potential energy from bonds ($\sum_{bond}$), bend angles ($\sum_{bend}$), and dihedral angles ($\sum_{dihedrals}$); the intermolecular interactions are usually accounted by Lennard-Jones$\left(\sum_{LJ}\right)$ and Coulomb ($\sum_{coulomb}$) [29].

$$k=\sum\nolimits_{bond}\frac{1}{2}k_{ij}(r-r_0)^2+\sum\nolimits_{bend}^{angle}\frac{1}{2}k_{ijk}(\theta-\theta_0)^2$$
$$+\sum\nolimits_{dihedral}k_{ijkl}(1+\cos(m\varphi-\delta))$$
$$+\sum\nolimits_{LJ}4\varepsilon_{ij}\left[\left(\frac{\sigma_{ij}}{r_{ij}}\right)^{12}-\left(\frac{\sigma_{ij}}{r_{ij}}\right)^{6}\right]+\sum\nolimits_{Coulomb}\frac{q_iq_j}{4\pi\varepsilon_0}\frac{1}{r_{ij}} \quad (6.59)$$

Eq. (6.59) calculates the atomic potential energy (k) using molecular dynamics; k_{ij}, k_{ijk}, and k_{ijkl} are the constants for bond, angle, and dihedral intramolecular interactions, respectively; the bond length is shown by r; the bond angle is indicated by θ; the dihedral angle is denoted by φ; atomic radii is symbolized by σ; ε is the well depth which refers to the local minimum of potential energy; and the charge of each atom is represented by q_i. The parameter δ adjusts the phase degree in the dihedral potential. Molecular dynamics can be used to estimate the equilibrium properties such as density or enthalpy of mixtures. This approach is proper to investigate transport properties such as the molecular coefficients that govern the rate of mass transfer (diffusion coefficients), heat transfer (thermal conductivity), and/or momentum transfer (viscosity). Molecular dynamics can be used to simulate physical properties such as interfacial tension, solubility, adsorption, and aggregation. However, it needs to be coupled with other modeling approaches such as quantum mechanics when there is a chemical reaction that produces a new component (molecule) in the system.

Typically, four statistical ensembles are considered in molecular dynamics, which can be implemented based on the test conditions, including (1) constant number of particles, volume, and energy (NVE), (2) constant number of particles, volume, and temperature (NVT), (3) constant number of particles, pressure, and temperature (NPT), and (4) constant chemical potential, volume, and temperature (μVT). The molecular dynamics method is developed to calculate the physical properties of the chemicals and their mixtures based on the chemical structures and molecular interactions (e.g., solubility and self-aggregation).

6.5 Smart models

In the past decade, smart models have found tremendous applications in different science and engineering disciplines, including processes in oil and gas research. Due to their high demand, a wide variety of the smart models are introduced in the literature. The smart models can be hybridized with first principle models to result in more effective and accurate predictions. To the best of our knowledge, four main groups of smart models are used in the literature to predict the asphaltene aggregation in the presence of chemical inhibitors. These groups are artificial neural network (ANN), least-squares support vector machine (LSSVM), group method of data handling (GMDH), and adaptive neuro-fuzzy inference system (ANFIS). Usually, these models are tuned using various optimization techniques, including coupled simulated annealing (CSA), genetic algorithm (GA), Bayesian regularization (BR), scaled conjugate gradient (SCG), ant colony optimization (ACO), Levenberg–Marquardt (LM), imperialistic competitive algorithm (ICA), conjugate gradient with Fletcher-Reeves updates (CGF), and particle swarm optimization (PSO).

6.6 Simulation tools

Some commercial software packages such as PVTsim (Calsep, Denmark), WINPROP (CMG, Canada), Multiflash (KBC, USA), and VLXE (VLXE ApS, Denmark) incorporate different models (or approaches) to predict asphaltene inhibition. For instance, PVTsim models the asphaltene deposition in the presence of various gas phases using PC-SAFT [30]. Multiflash simulates the asphaltene behavior using the SRK EoS and is capable of including new components [31]. The Flow Assurance Software Models is another software developed by AsphWax Inc., USA, to model the asphaltene inhibitor behavior [32]. Fig. 6.3 illustrates the main points in the macromodeling, which are used in the mentioned software. Moreover, molecular modeling can be applied with commercial software such as Materials Studio (BIOVIA, USA) [33], or open source software such as Gromacs [34]. Quantum mechanics modeling is also available as a toolbox in the Materials Studio [21, 22] and Gaussian 09 software [20]; this modeling approach is able to simulate the asphaltene-inhibitor interaction. Fig. 6.4 depicts the main steps that need to be implemented during the molecular modeling simulation in the mentioned software.

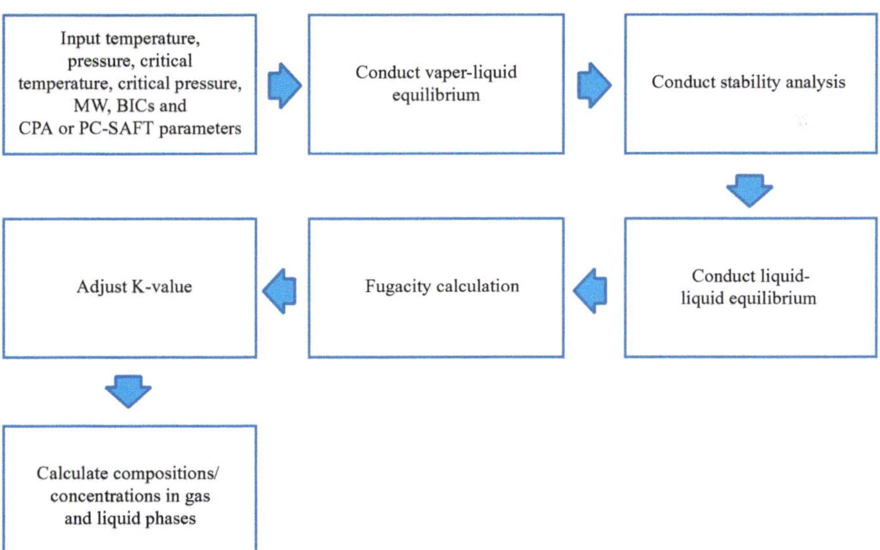

Fig. 6.3 The main points of three-phase macromodeling.

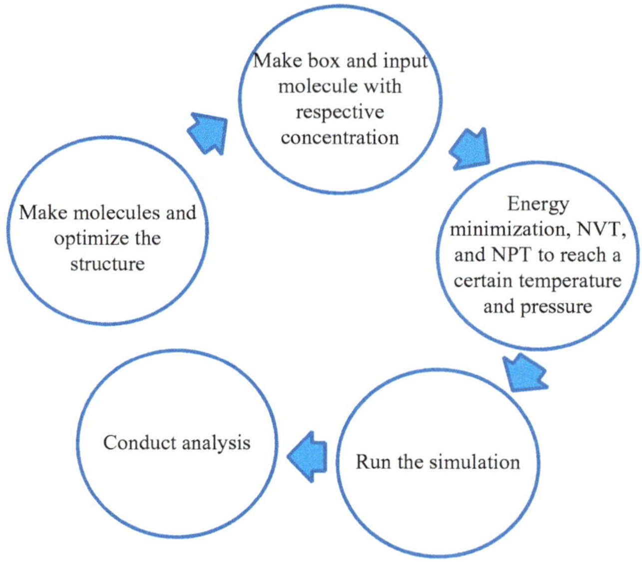

Fig. 6.4 The key steps of molecular modeling.

6.7 Parametric sensitivity analysis

Different modeling approaches can predict the asphaltene behavior in various processes with different operation conditions, fluid properties, and reservoir/pipe properties. In some studies, researchers compared the types of modeling methods and effect of time on inhibitor efficiency, which are discussed in the process/operational condition section. In the fluid properties section, the research studies that employed modeling methods to assess the effect of different inhibitors, asphaltenes, and oils are reviewed. To the best of our knowledge, there is no modeling study that investigates the sensitivity of the reservoir properties such as permeability in asphaltene deposition inhibition. The main reason is that the lack of experimental studies, which make the verification of model impossible.

6.7.1 Effects of process/operational conditions

Equations of state (EoSs), empirical equations, and thermodynamic methods can model the asphaltene precipitation and deposition with and without chemical inhibitors. Yonebayshi and Zhang [35, 36] used a CPA EoS based on SRK to obtain the asphaltene phase envelope (APE) with or without a commercial inhibitor. They introduced the inhibitor as a pseudo-resin component to model the asphaltene behaviors and compare the model outputs with the experimental results. Fig. 6.5 demonstrates the effect of the inhibitor (50 ppm) on APE under the operational conditions of asphaltene deposition as output of the model. One of the sources of error in the mathematical model was in using average molecular weight for the inhibitor instead of considering the molecular weight distribution [35, 36]. Assaf et al. [37] used PC-SAFT to predict the amount of deposition after the addition of dodecylbenzene sulfonic acid (DBSA). The liquid-liquid equilibrium was considered in the model, and the results were in agreement with the experimental filtration data [37]. Dehaghani and Badizad [38] coupled an association factor (called Dehaghani association) with the PR-EoS to consider the self-association of the asphaltene molecules and cross-association between asphaltene-inhibitor. They estimated the onset point of asphaltene precipitation in the presence of linear DBSA, branched DBSA, and cocamide diethanolamine inhibitors. The results showed a good match with experimental data, only at low concentrations of inhibitors. The model's error at high inhibitor concentration was caused by the colloidal behavior of the asphaltene in an experimental test, which contradicts the assumption of considering solubility theory in the EoSs [38]. Karambeigi et al. [39] used the acentric factor as a fitting parameter to match the modeling results for the

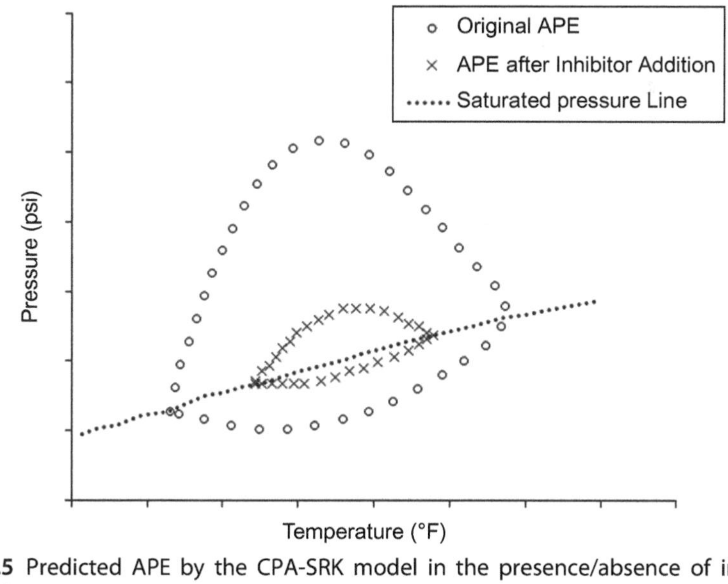

Fig. 6.5 Predicted APE by the CPA-SRK model in the presence/absence of inhibitor [35, 36].

asphaltene-inhibitor interactions, using experimental results. The reason for considering the acentric factor as a fitting parameter was its large impact on the APE [39].

Based on the adsorption type (single-layer adsorption or multilayer adsorption) and the type of surface (homogeneous or heterogeneous), either Langmuir or Freundlich isotherms are employed to model the asphaltene adsorption on nanoparticles accurately [40–43]. For instance, the Freundlich model was successful in modeling the asphaltene adsorption on silica gel nanoparticles coated with nickel oxide nanoparticle [40, 41], alumina nanoparticles with different acidity [43], fumed silica gel nanoparticles, alumina nanoparticles coated with low concentration nickel oxide, commercial aluminum (I), and zeolite [41]. The Langmuir model was successful in modeling the asphaltene adsorption on an alumina nanoparticles coated with high concentrations of nickel oxide, amorphic silica, PdNi/Al, and commercial silica gel [41], Fe_3O_4, Co_3O_4, TiO_2, MgO, CaO, and NiO (see Fig. 6.6) [42]. In 2017, Madhi et al. [44] used the Temkin and Dubinin-Radushkevich models along with the Langmuir and Freundlich models to study the adsorption of asphaltenes on silica, alumina, and magnesium oxide nanoparticles. They found both the multilayer adsorption and physisorption as the main mechanisms of asphaltene-nanoparticle adsorption and reported the Dubinin-Radushkevich model as the best EoS [44].

Barcenas and Duda [45] proposed a simple new model for particle agglomeration control (PAC) and used the canonical Monte Carlo method to study the cluster size distribution and the number of association sites between colloids as a function of inhibitor concentration. They reported

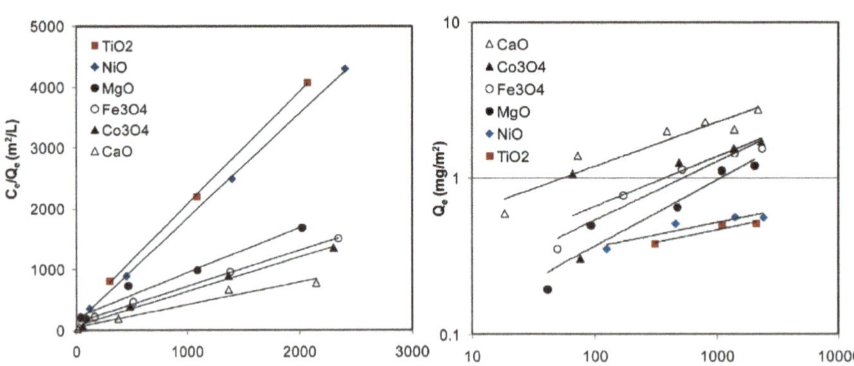

Fig. 6.6 Langmuir (left) and Freundlich (right) adsorption isotherm modeling of asphaltene adsorption on different nanoparticles [42].

the PAC model to be more suitable over alternative models using linear asphaltene aggregates.

The effectiveness of inhibitors at reservoir scale is important and should be included in the modeling studies. However, dynamic modeling at reservoir scale has not received much attention compared to static studies of the asphaltene-inhibitor interactions. Darabi and Sepehrnoori [10] tested DBSA treatment at the near wellbore for an oil well in the Middle East, using a mathematical model. After 200 operating days, the reservoir permeability was reduced to 70% of its initial value due to the asphaltene deposition; the oil production rate was 10% lower than the initial value. After treatment with 400 gal of DBSA and 2500 gal of water, the production profile of oil increased approximately 50 STBD compared to the case with no treatment. The soaking time and job frequency were optimized; increasing the soaking time from 6 to 24 h improved the inhibitor efficiency, while after 24 h the inhibitor efficiency was constant. They tested three optimal consecutive treatment cycles injection after 200, 400, and 700 days, which recovered 41% of the lost production due to the asphaltene deposition [10].

6.7.2 Effects of fluid properties

Different studies model the impact of fluid properties (such as different types of oils, inhibitors, and asphaltenes) on inhibitor effectiveness. Fundamental thermodynamic models minimize the interaction energy to find the most stable bonding interactions and to evaluate the efficiency of the inhibitor. For example, Pan and Firoozabadi [3] used the adsorption enthalpy (the adsorption heat for the asphaltene-inhibitor interactions) to evaluate the inhibitory effect of resin, dodecylphenol (DP), and DBSA in asphaltene deposition. They concluded that the adsorption enthalpy is higher for DBSA due to its higher polarity; consequently, DBSA is more effective than other inhibitors. The model did not consider the effect of self-association and the tail length of the inhibitors [3]. Victorov and Firoozabadi [46] introduced a thermodynamic model to predict the onset point of asphaltene precipitation. Later, Al Sahhaf et al. [2] modified the Victorov and Firoozabadi's model to simulate the asphaltene onset point in the presence of toluene, deasphalted oil, resin, nonylphenol, dodecyl resorcinol (DR), and DBSA. Their results showed that the model has less accuracy if the inhibitor exists only in the oil bulk and no inhibitor attaches to asphaltene shells. The accuracy becomes even less at a higher inhibitor concentration. Moreover, the inhibitors increased the resin concentration up to their saturation concentration (in the shell). For instance, a low concentration of DR increased the resin concentration up to 80% in the shell while it was limited to 50% when toluene was used as the inhibitor [2]. In 2008, Rogel [47] proposed a thermodynamic model to evaluate the effect of resin on asphaltene size and maximum soluble concentration of asphaltene. In agreement with the experimental results, the addition of resin decreased the asphaltene aggregate size and increased the maximum soluble concentration of the asphaltenes. Rogel [47] concluded that the solubility of resin polyaromatic rings in the bulk fluid decreases its inhibitory effect. Therefore, the aggregation originated by the attraction forces between the aromatic structures, and the repulsion forces of the asphaltene aggregates, and both precipitant and the inhibitor tails. In 2010, Rogel [48] modeled the effects of various parameters on the inhibitor efficiency, including: the number (and activity) of active sites, the number of asphaltene polyaromatic rings, the characteristics of the inhibitor tail, and the concentrations of asphaltene and inhibitors. The results showed that the solubility of the asphaltene polyaromatic rings and the number of active sites on the asphaltene molecules control the effectiveness of inhibitors. The number of asphaltenes per aggregates did not significantly affect the size of the asphaltene aggregates. The presence of inhibitors and the length of inhibitor tail induced additional steric effects. The model showed an optimum concentration of inhibitors, depending on the characteristics of systems, such as the inhibitor solubility and inhibitor self-aggregation. Fig. 6.7 shows the

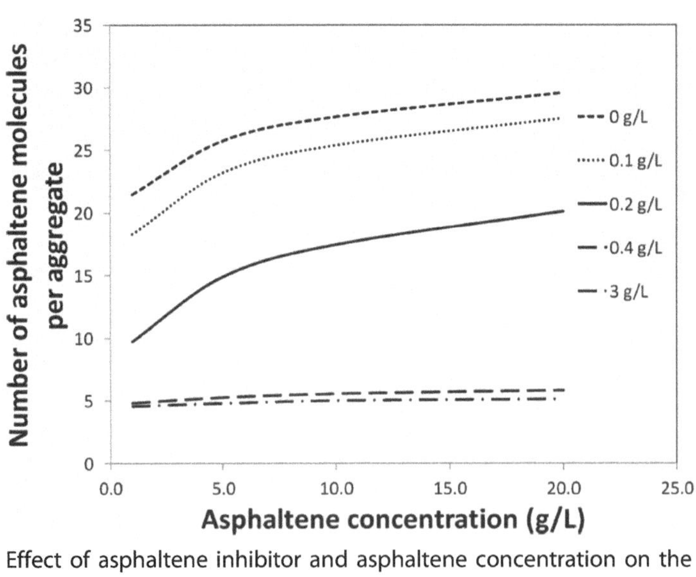

Fig. 6.7 Effect of asphaltene inhibitor and asphaltene concentration on the inhibitor efficiency [48].

effect of inhibitor concentration on the number of asphaltene molecules per aggregation versus the asphaltene concentration. Hydrogen bonding and acid-base interactions were not considered in the model; because of the dominant role of the aromatic interactions which were fair assumptions [48].

Molecular dynamics simulation is also used to study the inhibitor efficiency and inhibition mechanisms. Rogel and Leon [49, 50] investigated the behavior of inhibitors with different chemical structures. Aray et al. [51] and Sedghi et al. [52] used molecular dynamics to study different types of asphaltenes and inhibitors. Goual et al. [53, 54] studied the effectiveness of surfactants as asphaltene inhibitors. Lowry et al. [55, 56] studied the inhibitory mechanisms of polymer inhibitors. In 2001, Rogel and Leon [49] evaluated the interactions of dodecylphenol (DP), ethoxylated dodecylphenols and dodecyl benzene sulfonic acid (DBSA) with a high aromaticity asphaltene sample. They found the dipole moment and the polarity of inhibitors to control the interaction energy. The dipole moment and polarity were dominant when the asphaltene molecules were encapsulated with a low and high number of inhibitors, respectively. For low inhibitor concentration and at equilibrium condition, all the inhibitors had parallel configurations to the asphaltene surface except DBSA, which had a T-shape configuration. At a high inhibitor concentration, the heads of surfactant inhibitors were aligned parallel to the asphaltene surface and the tails extended isotopically toward the precipitant (*n*-heptane). Therefore, by increasing the chain size of inhibitor head, the atomic density close to the surface of the asphaltene molecules increased, restricting asphaltene flocculation. For this reason, ethoxylated alkylphenols had greater inhibitory effects than alkylphenols [49]. Rogel et al. [50] also studied the interactions of different inhibitors with asphaltene and confirmed the results of the previous studies, through examining various inhibitors such as: diethanol dodecylamine, *N,N*-dodecyl-2-pyrrolidone, *p*-(*n*-nonyl)phenol, *p*-(*n*-nonyl) phenol ethoxylated (6), *p*-(*n*-dodecyl)benzene sulfonic acid, and sorbitan monooleate [50]. In 2011, Aray et al. [51] studied nine islands and two archipelago asphaltene structures in toluene, *n*-heptane, and used surfactant inhibitors, including pyrrolidone, *p*-(*n*-nonyl)phenol, *p*-hexylphenol, *p*-(sec-butyl)phenol, *p*-(nonyl)phenol ethoxylated, Noram C, 6-methyl DBT, and two resins. The asphaltenes showed different structures, including the number of aromatic rings, heteroatoms, side chain, and the number of carbon atoms in the side chains. The surfactant inhibitors with no aromatic rings and long aliphatic tails or with aromatic rings surrounded by the aliphatic chains had high adsorption affinities on the asphaltenes with archipelago structures. Surfactant inhibitors with an aromatic ring and short aliphatic chains showed high adsorption affinities on the asphaltenes with island structures [51]. Goual et al. [53, 54] studied the interaction mechanisms between both octylphenol and DBSA with asphaltene, using the molecular dynamics simulation. They hypothesized an asphaltene island structure which formed acid-base and hydrogen bonding interactions with the inhibitors due to the presence of a pyridine group, a hydroxyl group, and a sulfide group. The Gibbs free energy for hydrogen bonding of the hydroxyl and pyridine functional groups (on the asphaltene), with inhibitor were −14 and −6.18 kJ/mol, respectively [53]. Therefore, octylphenol could interact easier with the asphaltene perimeter than its core, which prevented the asphaltenes lateral self-aggregation. Fig. 6.8 shows that early addition of octylphenol could prevent the asphaltene clustering and might even break the asphaltene nanoaggregates. However, the inhibitory effectiveness of octylphenol was uncertain, due to its self-aggregation potential, and its behavior as it approached (and bonds) to an asphaltene molecule [53]. In 2015, Goual et al. [54] compared the molecular behavior of DBSA and alkylphenols with asphaltene, and the inhibitory effect of DBSA was found inconclusive in different cases. The results showed that both DBSA and alkylphenol approach the perimeter of the asphaltene molecules, but with

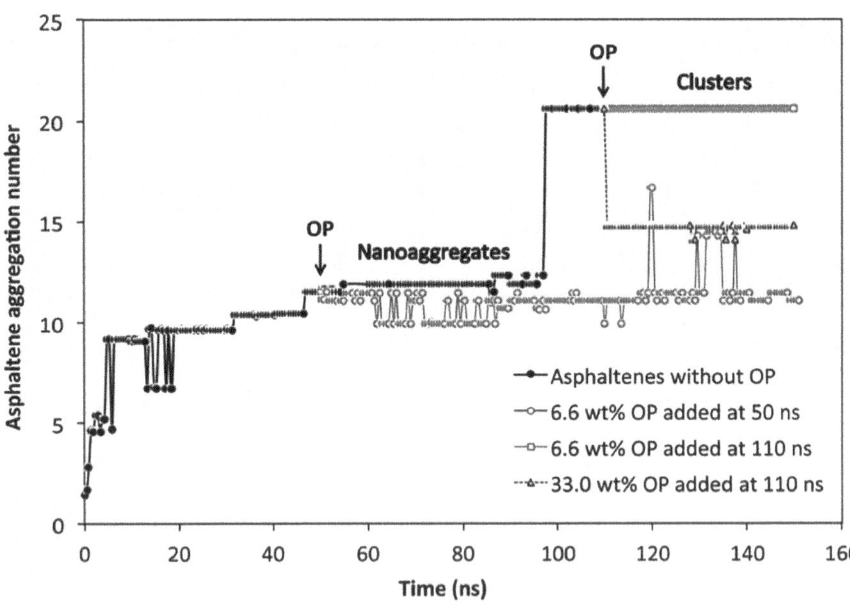

Fig. 6.8 Asphaltene aggregation number changes by time as a result of adding different concentrations of octylphenol to the solution of 6.6% asphaltene [53].

different mechanisms. The DBSA formed an electrostatic association with protonated asphaltene, which was 15 times stronger than hydrogen bonding energy that OP formed. The electrostatic association between the asphaltene-DBSA pairs caused their aggregation (the more favorably with end-to-end configuration), which did not occur for asphaltene-alkylphenol. The main bonding interaction of DBSA-asphaltene was acid-base, following hydrogen bonding.

The asphaltene-polymer interactions are also modeled with the molecular dynamics simulations. Lowry et al. [55] introduced a siloxane-based polymer, PDMS-g-propyl acetate (PPAc), as an effective, and environmentally friendly polymeric asphaltene inhibitor compared to poly(vinyl acetate) (PVAc) and poly(dimethylsiloxane) (PDMS). The interaction energies and the number of hydrogen bonds were measured between the polymeric inhibitors and three types of asphaltenes in a box mixed with supercritical carbon dioxide or near calcite surfaces surrounded by supercritical carbon dioxide. The molecular simulation of the inhibitors in the box showed PDMS to be ineffective as a dispersant. PVAc also lost its solubility after attaching to the asphaltenes due to the domination of van der Waals forces and electrostatic bonding. However, PPAc successfully reduced the aggregation onset of the asphaltenes in the simulation box; also, PPAc was even more effective for more polar asphaltenes. It showed stronger Coulomb interactions than PDMS with polar asphaltenes. Furthermore, PPAc formed fewer hydrogen bonds than PVAc due to the higher strengths of polar interactions, and higher density of carbonyl polar groups in the PVAc chain. When asphaltenes were adjacent to the calcite surface, they formed hydrogen bonds with the surface; however, the addition of PPAc polymer prevented bonding to the asphaltene surface [55]. In another study, Lowry et al. [56] studied the effectiveness of three polymeric inhibitors including M1: a maleic anhydride derived copolymer and two succinimide-based structures; namely one coupled with a phenol substituted oxazoladine group (S1), and one contained an interior alkene amine chain (S2). They studied three types of asphaltenes, including one without functional groups, one with an oxygen group, and one with a nitrogen group. All polymers effectively reduced the number of asphaltene aggregates, while both the asphaltene-M1 and asphaltene-S2 complexes showed self-association, which increased the aggregation (Fig. 6.9). Therefore, the S1 polymer was the most effective inhibitor due to its ability to interact asymmetrically with the asphaltene aggregates and strongly interact with the asphaltene aromatic cores. S1 also caused a larger repulsive force between the aggregates as

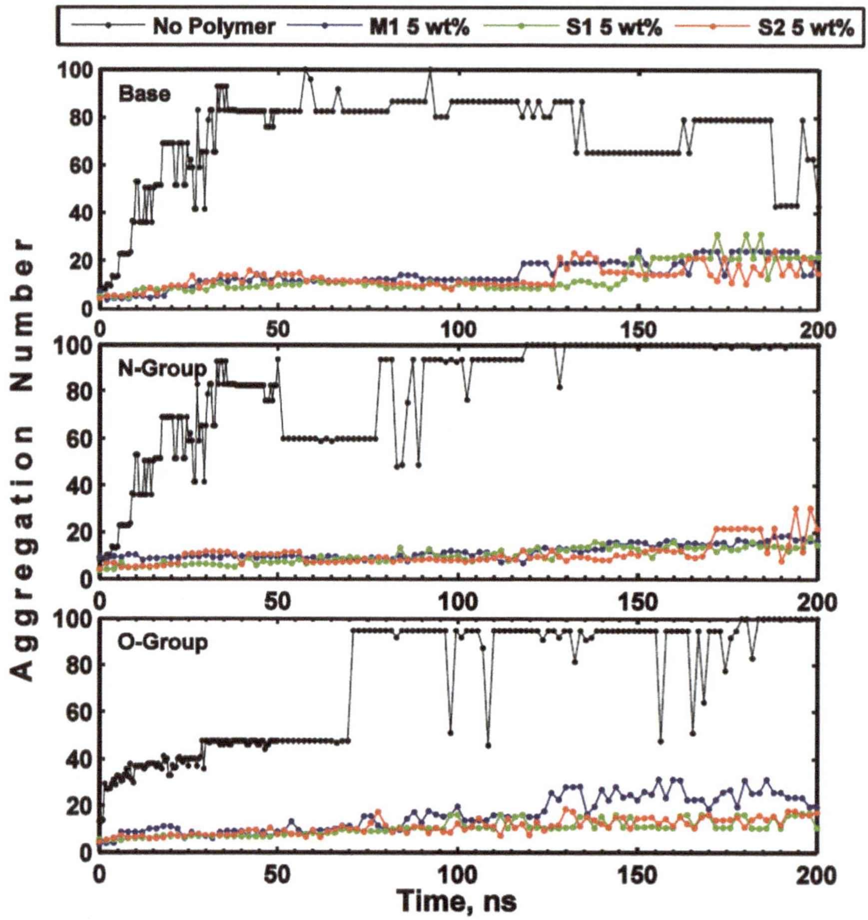

Fig. 6.9 Aggregation number of asphaltene as a function of time at 5% polymer [56].

a result of a surrounding poly(*iso*butylene) chain around the aromatic core (see Fig. 6.10). Although the S2 polymer caused positive Coulomb forces (repulsive forces) between asphaltene molecules, it created a semicross-linked structure due to the formation of intra- and intermolecular hydrogen bonds. The intramolecular interactions were related to the presence of secondary amine groups, un-coordinated oxygen atoms on the succinimide ring, and the intermolecular interactions because of the high density of the receptor and donor sites within the polymer chains. The M1 polymer bonded to the asphaltene molecules due to the interactions between styrene rings and the aromatic cores of the asphaltene. However, M1 featured the lowest solubility and the highest fraction of self-clusters among all tested

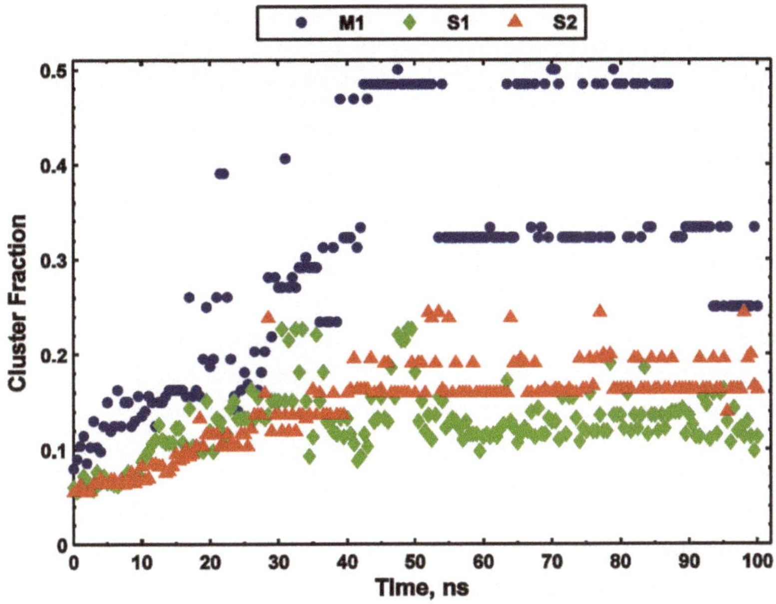

Fig. 6.10 The normalized average cluster size for each polymer type throughout the 100 ns solubility simulation in *n*-C5 [56].

inhibitors, which is because of the high Coulomb interaction energy (Fig. 6.10).

Molecular dynamics simulation is used to compare the effectiveness of two inhibitor types such as surfactants and polymers. Sedghi et al. [52] modeled the interactions for two different inhibitors, including limonene and PVAc, with two types of asphaltenes which have different polarities. They measured the interaction energies between the asphaltene-inhibitor aggregates, between the asphaltenes, and between the inhibitors. They also measured both the average number of asphaltene molecules in nano-aggregates and visualized the structural bonding of asphaltenes to evaluate the effectiveness of inhibitors. Limonene was tested at a concentration of 10%–30% with both asphaltenes; it was more efficient with the less polar asphaltene, and at higher limonene concentration. The PVAc was tested at the concentration range 1%–5%, and it was found more effective than limonene in reducing the aggregation of polar asphaltene. However, the asphaltene-PVAc aggregates had more self-association, which were controlled by adding 30% limonene to the asphaltene-PVAc complex. Most of the covered molecular dynamics studies were conducted by Gromacs software. In 2019, Tirjoo et al. [57] studied the inhibitory effects of several inhibitors, using Materials Studio software. They studied inhibitors, including linear DBSA, limonene, oleic acid, stearic acid, and two other inhibitors (diethylene triamine-penta methylene phosphonic acid and polyphosphino carboxylic acid). In their study, an asphaltene molecule with five aromatic rings and two methyl tails were used. Fig. 6.11 shows the radial distribution function for an asphaltene in the presence of a fixed number of different inhibitor molecules. The effectiveness of the inhibitors increased with increasing its concentration until an optimum level at which the inhibitors' self-aggregation became more probable than forming cross-association with asphaltenes. For all inhibitors, the optimum number was five molecules—except for limonene and stearic acid, which were three and ten molecules, respectively. Depending on the type of inhibitor, the asphaltenes form different structures: face-to-face dimers in the presence of limonene, linear DBSA, and polyphosphino carboxylic acid inhibitors; T-shaped dimers in the presence of oleic acid inhibitors; and monomer with stearic acid and diethylene triamine-penta methylene phosphonic acid (the most efficient). Ghamartale et al. [58] studied the effect of *n*-octylphenol on the aggregation of an archipelago asphaltene and two continental asphaltenes. They considered both singular and binary mixtures of asphaltenes at 1 bar and 300 K. The *n*-octylphenol could form both π-π and hydrogen bonds with the asphaltene that included an aromatic core and functional groups (such as hydroxyl and pyridine). In the case of continental asphaltene with no capability to form hydrogen bonds, the inhibitor could not reduce the

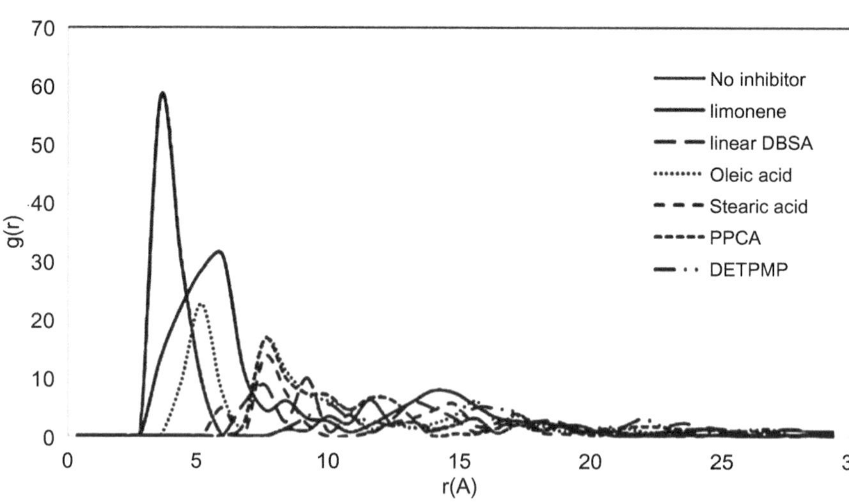

Fig. 6.11 Radial distribution function of the asphaltene molecule in the presence of different inhibitors [57].

aggregation. The *n*-octylphenol inhibitor was effective in reducing the asphaltene aggregation for the continental type asphaltenes with functional groups capable of forming hydrogen bonding.

The Monte Carlo simulation method is also used to study the effect of inhibitor type and concentration on asphaltene aggregation by estimating the probability of molecules aggregation. Barcenas et al. [59, 60] modeled two impermeable surfaces with several active sites as asphaltenes, and considered an inhibitor-solvent mixture between them to evaluate the asphaltene-inhibitor interactions. The results showed that inhibitors with a higher polarity have more affinity to adsorb onto the asphaltene surfaces at low concentrations; however, the inhibitors preferred to self-aggregate at a high inhibitor concentration. Fig. 6.12 shows the inhibitor behavior

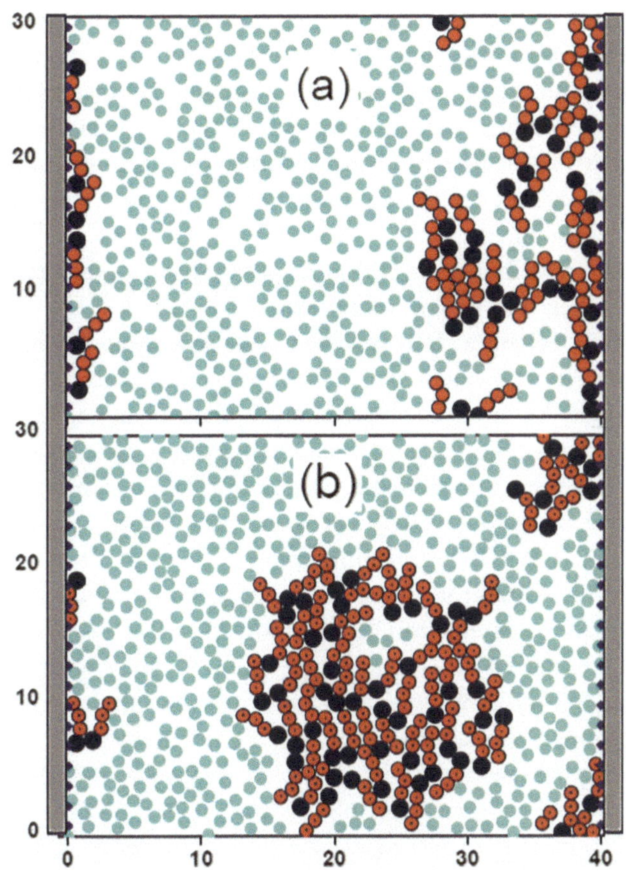

Fig. 6.12 Effect of inhibitor concentration on its preference to self-aggregate or approach to the asphaltenes: inhibitor concentration is 3:5 in a:b. Sidewalls are considered to be the asphaltene surface [60].

in different concentrations (number of inhibitor molecules per cell area is 0.031 and 0.5 in sections a and b, respectively). Barcenas and Orea [61] used the particle agglomeration control (PAC) theory and the canonical Monte Carlo method to capture the variations in molar-mass distributions (MMDs) in the presence of the nonylphenol and a commercial inhibitor. The inhibitor efficiency was affected by the asphaltene and inhibitor concentrations. In agreement with experimental data, increasing the number of inhibitor molecules increased the asphaltene aggregation (up to 11 asphaltene molecules) initially, but after reaching an extremum, the asphaltene aggregation started to decrease (to 7 asphaltene molecules). Barcenas and Duda [45] used the PAC theory and the canonical Monte Carlo method to evaluate the asphaltene cluster size with and without inhibitors. The asphaltene cluster size decreased upon the addition of the inhibitor, and the cluster size reduction was higher when the number of asphaltene connection site was increased. The maximum cluster size also decreased when the number of inhibitor molecules was increased [45]. Safaie and Nazar [62] studied asphaltene aggregation and deposition with dodecyl resorcinol and hazelnut oil, using the Monte Carlo simulation. Dodecyl resorcinol was found more effective than hazelnut oil; the hazelnut oil was effective only at high concentrations. Safaie and Nazar [62] were able to obtain an acceptable match for particle size distribution (PSD) and aggregation of asphaltene between simulation outputs and experimental data.

Quantum mechanics has received less attention in studying the asphaltene inhibition. Chavez et al. [22] used quantum mechanics and compared the interaction energies between the asphaltene aggregates, inhibitors, and the asphaltene-inhibitor complex. They demonstrated that the interaction energy between the aromatic poly(isobutenyl) succinimide inhibitor and asphaltene molecules was lower than the self-aggregation energy of the asphaltenes. The self-aggregation energy of all inhibitors was equal to or less than the inhibitor-asphaltene aggregation energy. Because the inhibitor-asphaltene interaction had the lowest interaction energy, it was found as the preferred interaction type [22]. In another study, Chavez et al. [21] used *n*-aryl amino alcohols with different aliphatic tail sizes and functional groups as asphaltene inhibitors. Three interactions were possible between the asphaltene and inhibitor molecules, including (1) ion–π interactions, due to the boronic moiety, (2) hydrogen bonds, due to the hydroxyl groups, and (3) parallel-displaced π-π stacking. The results showed that the inhibitor with boronic acid could form a tetrameric structure, which covered the aromatic section of the asphaltene properly. In addition, the inhibitor should have an aliphatic tail with enough length to form a stable steric site around the asphaltene. The most favorable inhibitor was the *n*-aryl amino alcohols which had both *n*-decan aliphatic chains and boronic acid functional groups. The interaction energy between the tetrameric structure inhibitors and asphaltene was -47 kcal mol^{-1}, which was lower than the asphaltene self-aggregation energy. The tetrameric structure could only interact with the asphaltenes from one side, and the other side repelled the asphaltenes, restricting the aggregation of the asphaltene-inhibitor complex [21]. Mosavi et al. [20] compared the asphaltene-resin interaction with the asphaltene self-aggregation, using density functional theory (DFT). They concluded that the number of aromatic fused rings in the asphaltene core dictates the number of resin molecules that are required to stabilize asphaltenes. The thermodynamic stability of the system increased when the number of resin molecules increased. As an example, increasing the number of resin molecules from one to three changed the binding energy from -17 to -47 kcal/mol; at a lower binding energy value, the asphaltenes were more stable. In the case of two resin molecules approaching from opposite directions toward an asphaltene molecule, lower interaction energy was reported compared to the approaching of resin molecules from one direction. Resin molecules were also found to stabilize the asphaltene dimers more efficiently than the asphaltene monomers, which was confirmed by less binding energy (-47 kcal/mol for the monomer and -60 kcal/mol for the dimer) [20].

The smart models are used to forecast the inhibitor influence on asphaltene aggregation based on available data. Baghban et al. [63] and Tashayo et al. [64] used multilayer perceptron (MLP) and radial basis function (RBF) as two ANN designs; they exhibited promising performance in predicting the asphaltene aggregation in the presence of chemical inhibitors. In addition, Malmir et al. [65], and Esfandiarian and Sedaghat [66] used ANFIS optimized with PSO and GA, and showed them as satisfactory tools in predicting the asphaltene aggregation behaviors while chemical inhibitors were included. The four mentioned studies used the same data bank that considered benzoic acid, phenanthrene, nonylphenol, phthalic acid, and salicylic acid as inhibitors. They accounted for model variables such as: the inhibitor structure, inhibitor concentration, and oil properties, including API, MW, asphaltene content, and the number of functional groups. In 2020, Mazloom et al. [67] conducted a comprehensive study on the prediction accuracy of different smart models in case of considering nanocomposites as asphaltene inhibitors. They considered effects of various variables and parameters on asphaltene aggregation, including solution pH, temperature, inhibitor-to-asphaltene ratio, and the inhibitor characteristics such as BET surface area and volume of micropores. They reported the following order in which the accuracy of models reduces: RBF-ACO, LSSVM-CSA, MLP-BR, MLP-LM, RBF-ICA, MLP-CGF, RBF-GA, RBF-PSO, and GMDH.

6.8 Asphaltene and inhibitor interaction coordination

The dominant interaction mechanisms that cause asphaltene aggregation and inhibition are still debatable in relevant studies. The preliminary idea is that the aromatic rings can have both attractive and repulsive forces, depending on their geometry. Fig. 6.13 shows three different mechanisms, including face-to-face, edge-on (or T-shape), and offset π-stacked, which are repulsive, attractive, and attractive, respectively. The asphaltene molecules are known for having either one (continental) or two (archipelago) polycyclic aromatic hydrocarbon cores. This feature enables them to attract to each other through induced dipole-induced dipole (due to resonance bonds) and dipole-dipole (due to heteroatoms) [68]; furthermore, they bond to each other with either of three mentioned coordinations (e.g., face-to-face, edge-on, and offset π-stacked). Headen et al. [69] and Teklebrhan et al. [70] studied the asphaltenes' coordination during aggregation; it was found that the offset stacking and T-shape stacking are the predominant mechanisms in dictating the coordinations of the asphaltene structure. Jian

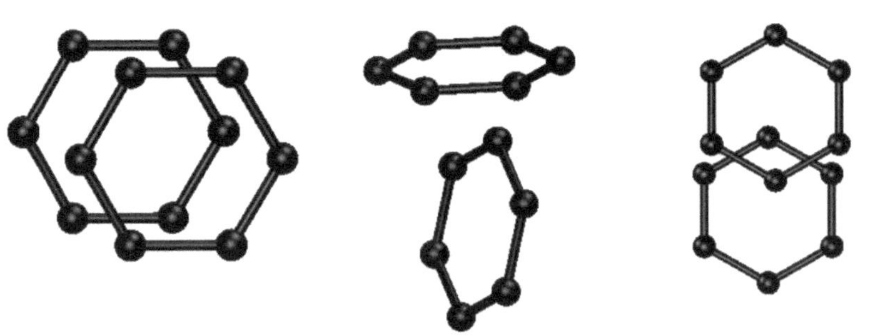

Fig. 6.13 Interactions between two aromatic rings based on their coordination.

et al. [71] stated that the dominant coordination is related to the asphaltene structure; they conducted a comprehensive study on the asphaltene aggregate coordination for asphaltenes with varying side-chain lengths. They proposed criteria for π-π stacking, including direct parallel stacking and shifted parallel stacking, π-θ stacking and θ-θ stacking [71]. Gray et al. [72] concluded that the π-π stacking is not the dominant force for asphaltene aggregation. They considered that Brønsted acid-base interactions, hydrogen bonding, metal coordination complexes, and interactions between cycloalkyl and alkyl groups (to form hydrophobic pockets) are also responsible asphaltene aggregation forces (Fig. 6.14). To the best of our knowledge, there are only a few limited studies on asphaltene-inhibitor interaction coordination. For instance, Goual et al. [53] and Ghamartale et al. [58] studied the asphaltene-inhibitor bonding and asphaltene aggregate shape in the presence of octylphenol; the importance of this matter necessitates more systematic studies to investigate the asphaltene-inhibitor coordination.

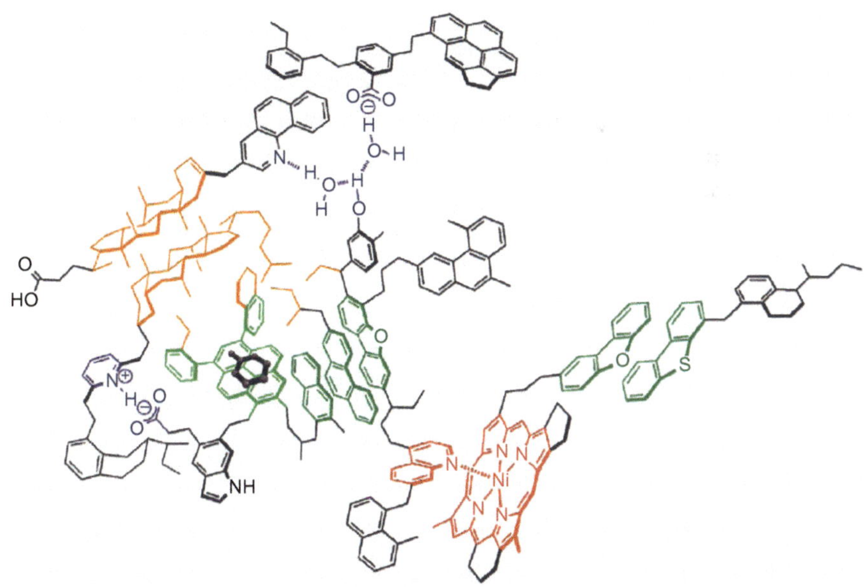

Fig. 6.14 Interactions between asphaltene molecules to form an asphaltene aggregate. The interaction bonds are color-coded: blue for acid–base and hydrogen bond interactions, red for metal coordination complex, orange for hydrophobic pocket, and green for π-π stacking [72].

6.9 Final remarks

This chapter reviewed different modeling approaches to study the asphaltene-inhibition behavior. The modeling approach requires complex computation, fitting or optimizing parameters, self- or cross-association parameters, understanding of the asphaltene structure, and different types of molecule's bondings. Despite all difficulties in modeling the asphaltene precipitation, deposition, and inhibition, some methods such as the EoSs with fitting parameters and molecular dynamics can describe the asphaltene-inhibitor behavior successfully. In addition, the LBM has great potentials in asphaltene modeling because of its advantage in recognizing the fluid-solid boundary and predicting nucleation. Therefore, LBM can be a great tool to model asphaltene-inhibitor behavior. The molecular scale modeling is a new approach to study the asphaltene-inhibitor interaction mechanisms. The asphaltene-inhibitor modeling in the dynamic condition is under development and needs future research. For instance, although researchers have focused on the asphaltene-asphaltene interactions and coordinations, there is no comprehensive study for asphaltene-inhibitor, to the best of our knowledge. Both the static and dynamic modeling of asphaltene-inhibitor interactions needs to be improved, which is the promising direction of future studies. In the advance modeling, the asphaltene-inhibitor interactions should be modeled in the presence of intruder (e.g., brine) and/or improver (e.g., resin) components.

References

[1] S. Subramanian, S. Simon, J. Sjöblom, Asphaltene precipitation models: a review, J. Dispers. Sci. Technol. 37 (7) (2016) 1027–1049.
[2] T.A. Al-Sahhaf, M.A. Fahim, A.S. Elkilani, Retardation of asphaltene precipitation by addition of toluene, resins, deasphalted oil and surfactants, Fluid Phase Equilib. 194 (2002) 1045–1057.
[3] H. Pan, A. Firoozabadi, Thermodynamic micellization model for asphaltene precipitation inhibition, AIChE J. 46 (2) (2000) 416–426.
[4] H. Pan, A. Firoozabadi, Thermodynamic micellization model for asphaltene aggregation and precipitation in petroleum fluids, SPE Prod. Facil. 13 (2) (1998) 118–127.
[5] B. Shirani, M. Nikazar, S.A. Mousavi-Dehghani, Prediction of asphaltene phase behavior in live oil with CPA equation of state, Fuel 97 (2012) 89–96.
[6] M. Sattari, J. Abedi, M. Zirrahi, A.K. Mehrotra, Modeling the onset of asphaltene precipitation in solvent-diluted bitumens using cubic-plus-association equation of state, in: SPE Canada Heavy Oil Technical Conference, Society of Petroleum Engineers, 2016.

[7] A. Arya, N. von Solms, G.M. Kontogeorgis, Determination of asphaltene onset conditions using the cubic plus association equation of state, Fluid Phase Equilib. 400 (2015) 8–19.
[8] J. Gross, G. Sadowski, Perturbed-chain SAFT: an equation of state based on a perturbation theory for chain molecules, Ind. Eng. Chem. Res. 40 (4) (2001) 1244–1260.
[9] J. Kondori, S. Zendehboudi, L. James, Evaluation of gas hydrate formation temperature for gas/water/salt/alcohol systems: utilization of extended UNIQUAC model and PC-SAFT equation of state, Ind. Eng. Chem. Res. 57 (41) (2018) 13833–13855.
[10] H. Darabi, K. Sepehrnoori, Modeling and simulation of near-wellbore asphaltene remediation using asphaltene dispersants, in: SPE Reservoir Simulation Symposium, Society of Petroleum Engineers, 2015.
[11] R. Miri, et al., Experimental and numerical modeling study of gravity drainage considering asphaltene deposition, Ind. Eng. Chem. Res. 53 (28) (2014) 11512–11526.
[12] M. Madhi, R. Kharrat, T. Hamoule, Screening of inhibitors for remediation of asphaltene deposits: experimental and modeling study, Petroleum 4 (2) (2018) 168–177.
[13] H. Liu, et al., Multiphase lattice Boltzmann simulations for porous media applications, Comput. Geosci. 20 (4) (2016) 777–805.
[14] M.B. Asadi, S. Zendehboudi, Hybridized method of pseudopotential lattice Boltzmann and cubic-plus-association equation of state assesses thermodynamic characteristics of associating fluids, Phys. Rev. E 100 (4) (2019) 043302.
[15] A. De Rosis, C. Coreixas, Multiphysics Flow Simulations Using D3Q19 Lattice Boltzmann Methods Based on Central Moments, arXiv preprint arXiv:2010.01628, 2020.
[16] H. Fazeli, M. Masoudi, R.A. Patel, P. Aagaard, H. Hellevang, Pore-scale modeling of nucleation and growth in porous media, ACS Earth Space Chem. 4 (2) (2020) 249–260.
[17] Y.-J. Lin, et al., Characterizing asphaltene deposition in the presence of chemical dispersants in porous media micromodels, Energy Fuel 31 (11) (2017) 11660–11668.
[18] P. Ungerer, V. Lachet, B. Tavitian, Applications of molecular simulation in oil and gas production and processing, Oil Gas Sci. Technol. 61 (3) (2006) 387–403.
[19] O.C. Mullins, et al., Advances in asphaltene science and the Yen–Mullins model, Energy Fuel 26 (7) (2012) 3986–4003.
[20] M. Mousavi, T. Abdollahi, F. Pahlavan, E.H. Fini, The influence of asphaltene-resin molecular interactions on the colloidal stability of crude oil, Fuel 183 (2016) 262–271.
[21] T.E. Chávez-Miyauchi, L.S. Zamudio-Rivera, V. Barba-López, E. Buenrostro-Gonzalez, J.M. Martínez-Magadán, N-aryl amino-alcohols as stabilizers of asphaltenes, Fuel 110 (2013) 302–309.
[22] T.S.E. Chávez-Miyauchi, L.S. Zamudio-Rivera, V. Barba-López, Aromatic polyisobutylene succinimides as viscosity reducers with asphaltene dispersion capability for heavy and extra-heavy crude oils, Energy Fuel 27 (4) (2013) 1994–2001.
[23] N. Metropolis, A.W. Rosenbluth, M.N. Rosenbluth, A.H. Teller, E. Teller, Equation of state calculations by fast computing machines, J. Chem. Phys. 21 (6) (1953) 1087–1092.
[24] A. Doucet, N. De Freitas, N. Gordon, An introduction to sequential Monte Carlo methods, in: Sequential Monte Carlo Methods in Practice, Springer, 2001, pp. 3–14.
[25] K. Liffman, A direct simulation Monte-Carlo method for cluster coagulation, J. Comput. Phys. 100 (1) (1992) 116–127.
[26] Y. Efendiev, M. Zachariah, Hybrid Monte Carlo method for simulation of two-component aerosol coagulation and phase segregation, J. Colloid Interface Sci. 249 (1) (2002) 30–43.
[27] A.L. Garcia, C. Van Den Broeck, M. Aertsens, R. Serneels, A Monte Carlo simulation of coagulation, Physica A 143 (3) (1987) 535–546.

[28] H. Zhao, C. Zheng, M. Xu, Multi-Monte Carlo method for coagulation and condensation/evaporation in dispersed systems, J. Colloid Interface Sci. 286 (1) (2005) 195–208.
[29] M.P. Allen, Introduction to molecular dynamics simulation, in: Computational Soft Matter: From Synthetic Polymers to Proteins, vol. 23, 2004, pp. 1–28.
[30] Footer Brochure, n.d. http://www.calsep.com/uploads/Modules/Footerbrochure/technical.pdf (accessed 20 November 2020).
[31] P. Bahrami, R. Kharrat, S. Mahdavi, H. Firoozinia, Prediction of the gas injection effect on the asphaltene phase envelope, Oil Gas Sci. Technol. 70 (6) (2015) 1075–1086.
[32] Simulators, n.d. http://www.asphwax.com/simulators.html (accessed 15 October 2020).
[33] R. Li, H. Du, Z. Fan, J. Pei, Molecular dynamics simulation to investigate the interaction of asphaltene and oxide in aggregate, Adv. Mater. Sci. Eng. 2016 (2016), 3817123.
[34] S. Yaseen, G.A. Mansoori, Molecular dynamics studies of interaction between asphaltenes and solvents, J. Petrol. Sci. Eng. 156 (2017) 118–124.
[35] H. Yonebayashi, X. Zhang, Emerging numerical modelling technique to evaluate asphaltene inhibitor efficiency during entire field life, in: SPE Asia Pacific Oil & Gas Conference and Exhibition, Society of Petroleum Engineers, 2016.
[36] H. Yonebayashi, Case study of asphaltene flow assurance risk mitigation-emerging numerical model to evaluate inhibitor efficiency, in: 78th EAGE Conference and Exhibition 2016, 2016.
[37] M. Assaf, M. Vafaie-Sefti, M. Shadman, A. Serajian, S. Ahmadi, Z. Taherian, Experimental and modeling asphaltene precipitation in presence of DBSA using PC-SAFT EOS, Pet. Sci. Technol. 34 (17–18) (2016) 1534–1541.
[38] A.H.S. Dehaghani, M.H. Badizad, Inhibiting asphaltene precipitation from Iranian crude oil using various dispersants: experimental investigation through viscometry and thermodynamic modelling, Fluid Phase Equilib. 442 (2017) 104–118.
[39] M.A. Karambeigi, M. Nikazar, R. Kharrat, A novel approach for asphaltene inhibitor modeling, Pet. Sci. Technol. 34 (3) (2016) 274–279.
[40] F.B. Cortés, J.M. Mejía, M.A. Ruiz, P. Benjumea, D.B. Riffel, Sorption of asphaltenes onto nanoparticles of nickel oxide supported on nanoparticulated silica gel, Energy Fuel 26 (3) (2012) 1725–1730.
[41] C.A. Franco, N.N. Nassar, M.A. Ruiz, P. Pereira-Almao, F.B. Cortés, Nanoparticles for inhibition of asphaltenes damage: adsorption study and displacement test on porous media, Energy Fuel 27 (6) (2013) 2899–2907.
[42] N.N. Nassar, A. Hassan, P. Pereira-Almao, Metal oxide nanoparticles for asphaltene adsorption and oxidation, Energy Fuel 25 (3) (2011) 1017–1023.
[43] N.N. Nassar, A. Hassan, P. Pereira-Almao, Effect of surface acidity and basicity of aluminas on asphaltene adsorption and oxidation, J. Colloid Interface Sci. 360 (1) (2011) 233–238.
[44] M. Madhi, A. Bemani, A. Daryasafar, M.R. Khosravi Nikou, Experimental and modeling studies of the effects of different nanoparticles on asphaltene adsorption, Pet. Sci. Technol. 35 (3) (2017) 242–248.
[45] M. Barcenas, Y. Duda, Irreversible colloidal agglomeration in presence of associative inhibitors: computer simulation study, Phys. Lett. A 365 (5) (2007) 454–457.
[46] A.I. Victorov, A. Firoozabadi, Thermodynamic micellizatin model of asphaltene precipitation from petroleum fluids, AIChE J. 42 (6) (1996) 1753–1764.
[47] E. Rogel, Molecular thermodynamic approach to the formation of mixed asphaltene – resin aggregates, Energy Fuel 22 (6) (2008) 3922–3929.
[48] E. Rogel, Effect of inhibitors on asphaltene aggregation: a theoretical framework, Energy Fuel 25 (2) (2010) 472–481.

[49] E. Rogel, O. León, Study of the adsorption of alkyl-benzene-derived amphiphiles on an asphaltene surface using molecular dynamics simulations, Energy Fuel 15 (5) (2001) 1077–1086.
[50] E. Rogel, E. Contreras, O. Leon, An experimental theoretical approach to the activity of amphiphiles as asphaltene stabilizers, Pet. Sci. Technol. 20 (7–8) (2002) 725–739.
[51] Y. Aray, R. Hernández-Bravo, J.G. Parra, J. Rodríguez, D.S. Coll, Exploring the structure–solubility relationship of asphaltene models in toluene, heptane, and amphiphiles using a molecular dynamic atomistic methodology, Chem. A Eur. J. 115 (42) (2011) 11495–11507.
[52] M. Sedghi, L. Goual, Molecular dynamics simulations of asphaltene dispersion by limonene and pvac polymer during CO_2 flooding, in: SPE International Conference and Exhibition on Formation Damage Control, Society of Petroleum Engineers, 2016.
[53] L. Goual, M. Sedghi, X. Wang, Z. Zhu, Asphaltene aggregation and impact of alkylphenols, Langmuir 30 (19) (2014) 5394–5403.
[54] L. Goual, M. Sedghi, Role of ion-pair interactions on asphaltene stabilization by alkylbenzenesulfonic acids, J. Colloid Interface Sci. 440 (2015) 23–31.
[55] E. Lowry, M. Sedghi, L. Goual, Novel dispersant for formation damage prevention in CO2: a molecular dynamics study, Energy Fuel 30 (9) (2016) 7187–7195.
[56] E. Lowry, M. Sedghi, L. Goual, Polymers for asphaltene dispersion: Interaction mechanisms and molecular design considerations, J. Mol. Liq. 230 (2017) 589–599.
[57] A. Tirjoo, B. Bayati, H. Rezaei, M. Rahmati, Molecular dynamics simulations of asphaltene aggregation under different conditions, J. Petrol. Sci. Eng. 177 (2019) 392–402.
[58] A. Ghamartale, S. Zendehboudi, N. Rezaei, New molecular insights into aggregation of pure and mixed asphaltenes in the presence of n-octylphenol inhibitor, Energy Fuel 34 (10) (2020) 13186–13207.
[59] M. Barcenas, P. Orea, E. Buenrostro-Gonzalez, L.S. Zamudio-Rivera, Y. Duda, Medium Dependence of Asphaltene Agglomeration Inhibitor Efficiency, arXiv preprint arXiv:0710.1289, 2007.
[60] M. Barcenas, P. Orea, E. Buenrostro-González, L.S. Zamudio-Rivera, Y. Duda, Study of medium effect on asphaltene agglomeration inhibitor efficiency, Energy Fuel 22 (3) (2008) 1917–1922.
[61] M. Barcenas, P. Orea, Molar-mass distributions of asphaltenes in the presence of inhibitors: experimental and computer calculations, Energy Fuel 25 (5) (2011) 2100–2108.
[62] K. Safaie, A.R.S. Nazar, Evaluation of asphaltene inhibitors effect on aggregation coupled sedimentation process, J. Dispers. Sci. Technol. 35 (3) (2014) 329–337.
[63] A. Baghban, R. Hekmati, M. Hajiali, M. Janghorban Lariche, M. Kamyab, Application of MLP-ANN as novel tool for estimation of effect of inhibitors on asphaltene precipitation reduction, Pet. Sci. Technol. 36 (16) (2018) 1272–1277.
[64] B. Tashayo, F. Zarei, H. Zarrabi, M.J. Lariche, A. Baghban, Utilization of RBF-ANN as a novel approach for estimation of asphaltene inhibition efficiency, Pet. Sci. Technol. 36 (16) (2018) 1216–1221.
[65] P. Malmir, M. Suleymani, A. Bemani, Application of ANFIS-PSO as a novel method to estimate effect of inhibitors on Asphaltene precipitation, Pet. Sci. Technol. 36 (8) (2018) 597–603.
[66] A. Esfandiarian, M. Sedaghat, Development of a novel ANFIS-GA method for investigation of effect of inhibitors on asphaltene precipitation, Pet. Sci. Technol. 37 (14) (2019) 1626–1633.
[67] M.S. Mazloom, F. Rezaei, A. Hemmati-Sarapardeh, M.M. Husein, S. Zendehboudi, A. Bemani, Artificial intelligence based methods for asphaltenes adsorption by nanocomposites: application of group method of data handling, least squares support vector machine, and artificial neural networks, Nanomaterials 10 (5) (2020) 890.

[68] O.C. Mullins, The modified Yen model, Energy Fuel 24 (4) (2010) 2179–2207.
[69] T.F. Headen, E.S. Boek, N.T. Skipper, Evidence for asphaltene nanoaggregation in toluene and heptane from molecular dynamics simulations, Energy Fuel 23 (3) (2009) 1220–1229.
[70] R.B. Teklebrhan, L. Ge, S. Bhattacharjee, Z. Xu, J. Sjöblom, Probing structure–nanoaggregation relations of polyaromatic surfactants: a molecular dynamics simulation and dynamic light scattering study, J. Phys. Chem. B 116 (20) (2012) 5907–5918.
[71] C. Jian, T. Tang, S. Bhattacharjee, Probing the effect of side-chain length on the aggregation of a model asphaltene using molecular dynamics simulations, Energy Fuel 27 (4) (2013) 2057–2067.
[72] M.R. Gray, R.R. Tykwinski, J.M. Stryker, X. Tan, Supramolecular assembly model for aggregation of petroleum asphaltenes, Energy Fuel 25 (7) (2011) 3125–3134.

CHAPTER SEVEN

Inhibitor screening workflow

7.1 Critical parameters for screening inhibitors

Based on the technical discussions in Chapters 4–6, there are various static and dynamic parameters related to the process types, operation conditions, asphaltene type, asphaltene medium, and inhibitors that are important in the selection of an effective inhibitor. The operating parameters include pressure and temperature, and the process parameters include the test type (static or dynamic tests) and the test design (simultaneous or sequential injection). The asphaltene parameters include the asphaltene polarity, the asphaltene acidity, the type and location of a functional group on asphaltene molecules, aromaticity (H/C ration), and the asphaltene concentration. The parameters related to the hydrocarbon medium include the acidity/basicity of fluid, SARA fraction, the type of solvent/precipitant in synthetic oil, and the water/brine presence. The rock or pipe surface characteristics, such as rock texture and pipe material, also play an important role on the effectiveness of inhibitors in the case of fluid flow in a porous medium or a pipe. The inhibitor parameters depend on inhibitor types including surfactants, polymers, nanoparticles, ionic liquids, and organic inhibitors. All mentioned parameters are important in the screening of inhibitors and including more parameters leads to more reliable results. We divide these influential parameters into general screening criteria and inhibitor-specific screening criteria. The general parameters can change the effectiveness of all types of inhibitors; however, the inhibitor-specific parameters depend on the inhibitor type. For instance, we consider temperature and environmental criteria as general parameters for all types of inhibitors. However, the acidity of the asphaltene and hydrocarbon phase greatly affect the effectiveness of ionic inhibitors, specifically, but not the effectiveness of nonionic inhibitors. Tables 7.1 and 7.2 summarize the general screening criteria, and inhibitor-specific screening criteria for each type of inhibitor, respectively.

Asphaltene Deposition Control by Chemical Inhibitors
https://doi.org/10.1016/B978-0-323-90510-7.00005-7

Table 7.1 Influential parameters on inhibition mechanisms: general parameters.

Influential general parameter		
Pressure and temperature	Process design	Asphaltene type
Oil type	Precipitant type	Surface type
Feasibility of implementing	Cost	Environmental side effects

Table 7.2 Influential parameters on inhibition mechanisms: dependent parameters.

Influential dependent parameters

Surfactants	Polymers	Nanoparticles	Ionic liquids	Oil products
Type and number of functional groups	Polymerization degree	Surface chemistry and charge	Functional groups	SARA ratio
Size and number of alkyl tails	Type and number of functional groups	Specific surface	Size of alkyl tail	H/C
Acidity	Molecular weight	Shape and size	Charge density	API Polarity Solubility

7.2 Technical screening criteria

7.2.1 Workflows

Researchers have suggested various workflows to determine the most effective inhibitor. Selecting an optimal inhibitor based on technical, practical, economical, and environmental perspectives has always been a challenge in the oil and gas industry. Thus far, the suggested workflows only rank the inhibitors based on technical performance and lack a generalization capacity. For instance, Albanny et al. [1] proposed a workflow for two types of asphaltene inhibitor treatment methods, including continuous downhole injection and soaking cycle injection. The performances of these methods were evaluated using an asphaltene dispersion test (ADT) under the atmospheric conditions. The compatibilities of the inhibitors with rock and reservoir fluids were evaluated. In the soaking cycle injection method, the inhibitor efficiency retention time was also examined. The workflow lacked dynamic tests (for continuous downhole inhibitor injection), and realistic pressure and temperature (for both treatment methods). The workflow also did not consider economic evaluation [1]. Buriro et al. [2] proposed an efficient workflow to select asphaltene inhibitors. The solid detection system, polarized light microscopy, and titration technique were used as static tests, while coreflood tests were conducted as dynamic tests at ambient and reservoir thermodynamic conditions. They considered five main screening criteria to selectthe efficient inhibitor including: (1) effective functional group, (2) effective tail length, (3) polarity, (4) optimum concentration, and (5) environmental impacts. The tests were, however, conducted using dead crude oil. In addition, important parameters, such as the repetition time, feasibility, and cost, were not considered in the workflow. Tavakkoli et al. [3] suggested an indirect procedure to define the asphaltene precipitation onset point and the asphaltene precipitation amount. Based on this procedure, samples were prepared with various percentages of precipitant (*n*-heptane) and were shaken by hand. After giving the samples the same resting time, they were centrifugated (15 min at 10,000 rpm) to separate the asphaltene particles larger than 100 nm. The supernatant fluid obtained from the samples was diluted with toluene at a volumetric ratio of 1:4 and prepared for spectrophotometry test. The light absorbance for diluted supernatants was investigated with spectrophotometer at a range of wavelengths (1300 nm backwards to 700 nm). The precipitant volume, in which the light absorbance drops suddenly, was considered as the starting point for asphaltene deposition. Bae et al. [4] proposed comprehensive technical screening workflows in 2016 where the asphaltene dispersion tests were performed at both atmospheric and elevated temperatures, using an optical scanning technique. As only a fraction of the precipitated asphaltenes contributes to the deposition process, the coupon deposition test was performed. A high temperature/pressure rocking cell was used to test the effectiveness of the inhibitors at elevated pressures and temperatures. In both the coupon deposition test and the high temperature/pressure rocking cell test, the effect of water on inhibitor efficiency was also tested. The workflow can be improved by including economic aspects and considering porous media in case of injecting inhibitor near wellbore [4].

Some studies use a simplified methodology to select an asphaltene inhibitor. For instance, Yonbayashi et al. [5] used ADT for inhibitor screening, where the most effective inhibitor was selected by comparing different inhibitors at a fixed concentration. The optimum concentration of the selected inhibitor was then determined by testing the inhibitor effectiveness in a range of concentrations. This procedure did not include dynamic effects, and high pressure and temperature conditions. Villard et al. [6] introduced a workflow using ADT and coreflooding tests to identify the most effective type and concentration of inhibitors under static and dynamic conditions. The compatibility between the inhibitor and oil was also tested. The high pressure and temperature conditions, and economic and environmental evaluation were not included in this workflow. Ortiz et al. [7] considered the economic factor in their workflow for inhibitor selection. They proposed a treatment workflow in the asphaltene deposition problem. In the workflow, the effective inhibitor type and optimal concentration were selected using the ADT test. Then, the injection strategy and the treatment fluid volume were chosen, and an economic analysis was conducted. The proposed workflow did not consider high pressure and temperature conditions, dynamic effects, and the inhibitor compatibility with the reservoir fluids.

All the screening methods that were discussed in this section are obtained based on experiments. It is also possible to conduct inhibitor screening using mathematical modeling and simulation. After the simulation results are validated with the experimental data, the model can be used in parametric sensitivity analysis to estimate the inhibitor performance under different conditions. Molecular dynamics simulation is a first principle model that is capable of studying the inhibitor-asphaltene interactions at molecular level under static conditions (at different pressures, temperatures, and chemical compositions). Rogel [8] conducted a comprehensive study based on molecular dynamics simulation to obtain the most effective inhibitor for asphaltene deposition mitigation. In this study, the molecular structures and the concentrations of both asphaltenes and inhibitors and solvent-related parameters were considered in the screening. Employing cubic EoSs is also another promising modeling method to conduct a comprehensive sensitivity analysis after being validated by experimental results.

Thus far, we summarized the general workflow for inhibitor screening according to the literature studies. A comprehensive screening method includes different types of tests (static and dynamic), in a desired range of pressure and temperature conditions to select an efficient, economical, and environmentally friendly inhibitor with an optimal concentration. In the following, we propose our instruction to screen asphaltene inhibitors (see Fig. 7.1):

- *Inhibitor type selection:* The type of asphaltene inhibitor is selected in this part. In Section 7.2.2, few tips are provided for a proper selection of the inhibitor type based on the oil and asphaltene properties.
- *Inhibitor prescreening:* Several inhibitors are tested using simple static tests and based on trivial factors such as availability, cost, and environmental concerns. ADT is one of the simplest, fastest, cheapest, and most popular

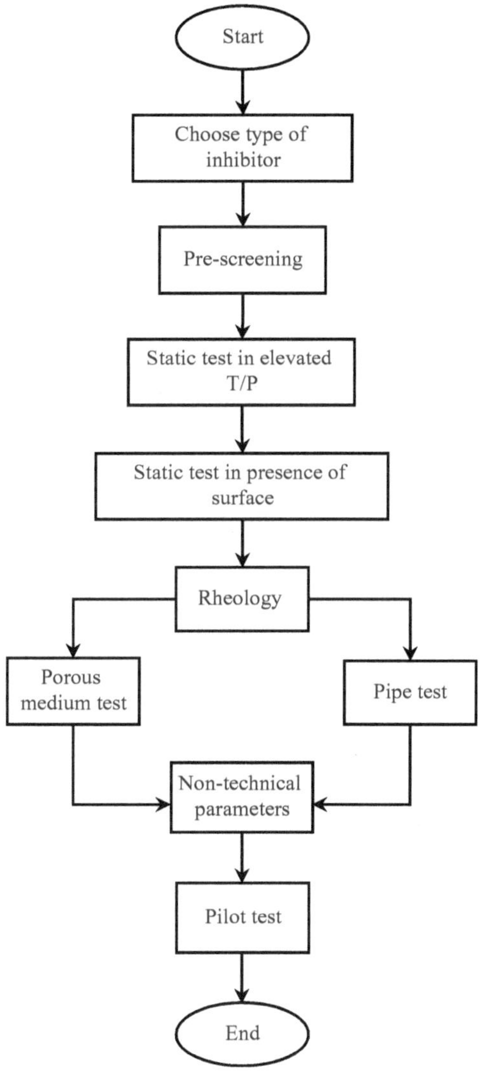

Fig. 7.1 The procedure of screening an asphaltene inhibitor.

methods among the static tests. Despite its simplicity and uncertainties in estimating the realistic efficiency of the inhibitors, it is a good prescreening measure.

- *Static tests at elevated pressures and temperatures:* Based on a study carried out by Melendez-Alvarez [9], the indirect method of asphaltene precipitation is the most effective strategy to assess the performance of an inhibitor for a target crude oil and reservoir operating conditions. However, the indirect method requires centrifugation, which is expensive if not impossible. The direct methods of asphaltene inhibition assessment are used when indirect methods are not possible; but the next step is necessary if the direct method is employed.

- *Static deposition tests:* When conducting the indirect method of asphaltene precipitation detection is not possible; a coupon deposition test will become necessary to find the effective inhibitor type and its optimal concentration using realistic measurements of asphaltene deposition. The tests can also be conducted to confirm the indirect test result. However, in the coupon deposition tests, the surface is (usually) stainless steel, and the pressure and temperature are atmospheric. These conditions may be modified for a more realistic assessment of the inhibitors.
- *Visual methods:* Field-emission scanning electron microscopy (SEM) and high-resolution transmission electron microscopy (TEM) are commonly used as visual methods to assess the efficiency of asphaltene inhibitors. However, the visual methods are not recommended as main assessment methods due to their operator-dependent nature. They can be utilized to verify the results of previous experiments, which make this step optional.
- *Rheology test:* Viscosity measurement is one of the common rheology tests that is categorized in dynamic tests because it can study the effect of shear. This test can be used to measure the onset point of asphaltene precipitation/deposition, using inhibitors and at different temperatures.
- *Dynamic test:* The dynamic tests vary based on the treatment type. In near wellbore treatment, where the effect of porous media becomes important, it is suggested to conduct the dynamic tests in porous media; otherwise, simple dynamic tests in pipes may be sufficient.
- *Dynamic tests in a porous medium:* Preferably, coreflooding tests using the formation core and live oil and at the reservoir condition are conducted. An outcrop core, a synthesis core, or a packed bed (with a similar material type to the actual rock, such as using glass to simulate the sandstone reservoir) are acceptable when the formation core is not available. The synthetic core should have a similar texture, permeability, and porosity with those of the formation core. If live oil is not available, dead oil or diluted crude oil can be used at reservoir temperature and not less than 30% crude oil in the case of using diluted crude oil [10]. If the temperature cannot be adjusted at the desired level, depending on the effect of temperature at the static condition, a dynamic test in room temperature is acceptable.

- *Dynamic tests in pipes:* There are extensive reports on asphaltene deposition tests using flow loops. These tests can also investigate the asphaltene inhibitor performance in pipes. Parameters such as flow rate, flow regime, pressure, and temperature are important in these tests. The inhibitor effectiveness can be measured by recording changes in pressure drop or pipe's inner diameter.
- *Feasibility at the field scale:* These tests provide the most realistic assessments of the asphaltene inhibition performance in terms of operational, financial, and environmental factors at the field scale. There are a few notes to consider
  - Note 1: If synthetic oil is used during the assessment tests, the results should either be confirmed with the crude oil tests or be repeated using different precipitants and solvents.
  - Note 2: Compatibility of the additive with reservoir fluid should be checked under operation conditions.
  - Note 3: The repeatability of the tests should be confirmed through replication tests.

7.2.2 Selecting the inhibitor type

Selecting the effective inhibitor type facilitates the screening process; however, there is no comprehensive study that provides recommendations on the selection guide for the inhibitor type based on the crude oil, asphaltene properties, and the operating conditions. Nevertheless, there are some heuristic criteria to select an efficient inhibitor type. Usually, more than one type of inhibitor may be selected for a specific case, and the inhibitors from all selected groups should then enter the workflow. The main screening parameters are temperature, oil viscosity, oil acidity, and asphaltene polarity. Based on the type of inhibitors, changing temperature can increase or decrease their efficiency. Surfactants, polymers, and ionic liquids become more effective when temperature increases, in case they do not decompose. However, nanoparticles are often more effective at low temperatures because asphaltene-nanoparticle interaction is an exothermic reaction. The oil viscosity is an important parameter in operations that require fluid injection into wells or reservoirs because it changes the fluid mobility and energy demand. To inhibit asphaltene precipitation from heavy oils at field scale, high-viscosity inhibitors such as polymers may not be favorable compared to the organic solvents. Sometimes due to viscosity and mobility changes, even if organic solvents have a lower nominal inhibitory efficiency in lab tests, they can outperform the polymeric inhibitors with higher effectiveness. The crude oil acidity and the asphaltene polarity limit the applicability of the inhibitors in terms of charge or acidity. The effective interaction mechanisms between ionic inhibitors and asphaltene are of the acid-base type; therefore, an inhibitor with a suitable charge should be selected. The selected inhibitor may be neutralized with the intruder components in the crude oil medium such as hydrocarbons or brine before interacting with the asphaltenes. In this case, nonionic inhibitors are suggested. For example, nonylphenol or ethoxylated nonylphenol with an optimized degree of ethoxylation can be effective alternatives.

7.3 Nontechnical screening criteria

Large-scale feasibility, economic, and environmental impacts are three important concerns in conducting pilot treatment in the field scale. In the following, each concern is investigated with more details.

7.3.1 Feasibility

The feasibility of treatment at the field scale is one of the first concerns to implement a designed treatment using the selected inhibitor. An inhibitor may be effective at the lab scale, but not at the field scale. For instance, the treatment in horizontal wells and multilateral wells requires more precautions, complex procedures, demanding a higher technology compared to vertical wells. In addition, the inhibitor preparation at large scale needs specialized knowledge and efforts; for instance, preparing a large amount of polymers or nanofluids requires special procedures and equipment. To conduct a treatment in a small/compact structure such as an offshore platform, space (and weight, specifically in offshore platforms) is a limiting factor. For example, nanofluid preparation requires special surface facilities, and they need to be prepared near the injection site; otherwise, the nanofluid stability will become an issue. Weather conditions can also become a concern based on the type of inhibitor. For instance, polymers may become viscous in very cold weather; similarly, solvents may evaporate in very hot weather. Therefore, the availability (and maturity) of the technology besides considering geographical situation, such as space and weather, are important concerns in the field-scale implementations.

7.3.2 Economic measures

Before any field operation, an economic analysis should be conducted to evaluate the cost-effectiveness. The economic analysis usually includes capital expenditures (CAPEX) and operating expenditures (OPEX) such as the costs of surface facilities, and costs of energy and materials, respectively. Di Lullo et al. [11], Cenegy [12], and Akbarzadeh et al. [13] conducted financial evaluations on the common asphaltene treatment methods, such as a solvent wash and scraper, and inhibitor treatments. They considered variables such as production rate, treatment frequency (and duration), and the improvement changes by treatment repetition. It was concluded that the inhibitor treatment is a cost-effective method compared to the common asphaltene treatment methods. A similar procedure can be employed to evaluate the economic perspective of the treatments with different inhibitors. The economic aspect is a major factor for the industry; if an inhibitor with high efficiency in lab scale cannot increase the production rate to compensate the treatment cost and make expected profit, the additional expenses are not economically justifiable [11]. The inhibitor cost is one of the variables in screening inhibitor treatment methods for mitigation/management of the asphaltene problems. In Chapter 3, we discussed about the cost and required concentration of each inhibitor. Table 7.3 briefly summarizes the minimum, maximum, and average inhibitors cost per unit for different types of inhibitors.

7.3.3 Environmental impacts

Societies are becoming continuously more concerned about the emissions of greenhouse gases from the oil and gas industry and the chemical treatments increase the risk of environmental and ecosystem pollution both in onshore and offshore reservoirs. One of the main concerns is the leakage of the chemical fluids in the undesired layer or surface, leading to the aquifers and ecosystem contamination [14]. Although the chemical inhibitors are more efficient compared to treatment methods such as scraping, the chemical methods usage increases the rick of environmental footprints and human safety. Therefore, an extra precaution and safety factor should be considered during the chemical treatment, causing a compensable damage to the ecosystem. The organic inhibitors and vegetable oils are safer alternatives to the chemical inhibitors to reduce the mentioned concerns; however, they feature a lower efficiency compared to the chemical inhibitors [15]. Three steps are taken to solve the environmental concerns by developing:

Table 7.3 The minimum, maximum, and average cost of using each type of inhibitor with the respective inhibitor in that category (there is not enough information about the polymer cost).

Type	US$ per	Min	Max	Average (without considering out layer)
Surfactants	kg	1 US$/kg (linear alkylbenzenesulfonic acid and sodium lauryl ether sulfate)	27,600 US$/kg (octylphenol)	288.95 US$/kg (sec-butylphenol)
	L	12.775 US$/L (oleic acid)	510 US$/L (2-methoxy-4-methylphenol)	77.88 US$/L (diethylamine)
Nanoparticles	kg	134.4 US$/kg (fumed silica gel)	21,600 US$/kg ($TiO_2$)	2910 US$/kg (CaO)
Solvents	kg	46.6 US$/kg (naphthalene)	2430 US$/kg (benzothiophene)	769.4 US$/kg (phenanthrene)
	L	30.9 US$/kg (toluene)	182 US$/kg (pyridine)	84.98 US$/kg (xylene)
Ionic liquids	kg	429 US$/kg (1-ethyl-3-methylimidazolium chloride)	15,160 US$/kg (1-butyl-3-methylimidazolium nitrate)	2717 US$/kg (1-octyl-3-methylimidazolium chloride)

(1) biodegradable asphaltene inhibitors such as bacteria [16]; (2) environmentally friendly chemicals such as polymers [17]; and (3) physical methods such as ultra-sonication [14].

7.4 Final remarks

This chapter covers the available screening criteria for asphaltene inhibitors based on previous research studies. Then, a comprehensive workflow was proposed based on the understandings from previous studies in the literature. The proposed workflow starts with selecting the type of inhibitor based on temperature, and characteristics of asphaltene and asphaltene medium. It is mentioned that the selected inhibitors should be assessed using static and dynamic tests at ambient and reservoir pressures and temperatures. The last section of the workflow focuses on the nontechnical parameters, which are also important factors in field-scale implementations. The treatment feasibility, economy, and environmental impacts are discussed as the nontechnical screening criteria.

References

[1] A. Albannay, T. Masuzawa, C.A. Dabbouk, Alternative methods in preventing asphaltene deposition, in: Abu Dhabi International Petroleum Exhibition and Conference, Society of Petroleum Engineers, 2010.

[2] M.A. Buriro, M.T. Shuker, Asphaltene prediction and prevention: a strategy to control asphaltene precipitation, in: SPE/PAPG Annual Technical Conference, Society of Petroleum Engineers, 2012.

[3] M. Tavakkoli, et al., Indirect method: a novel technique for experimental determination of asphaltene precipitation, Energy Fuel 29 (5) (2015) 2890–2900.

[4] J. Bae, D. Fouchard, S. Garner, J. Macias, Advantages of applying a multifaceted approach to asphaltene inhibitor selection, in: Offshore Technology Conference, Offshore Technology Conference, 2016.

[5] H. Yonebayashi, X. Zhang, Emerging numerical modelling technique to evaluate asphaltene inhibitor efficiency during entire field life, in: SPE Asia Pacific Oil & Gas Conference and Exhibition, Society of Petroleum Engineers, 2016.

[6] Y. Villard, F. Fajardo, A. Milne, Enhanced oil recovery using innovative asphaltene inhibitors in east venezuela, in: SPE International Conference and Exhibition on Formation Damage Control, Society of Petroleum Engineers, 2016.

[7] R. Ortiz, et al., Asphaltene prevention workflow enhances oil production in ht fractured carbonate reservoirs, in: SPE International Conference and Exhibition on Formation Damage Control, Society of Petroleum Engineers, 2016.

[8] E. Rogel, Effect of inhibitors on asphaltene aggregation: a theoretical framework, Energy Fuel 25 (2) (2010) 472–481.

[9] A.A. Melendez-Alvarez, et al., On the evaluation of the performance of asphaltene dispersants, Fuel 179 (2016) 210–220.

[10] D.W. Jennings, K.-P. Chao, J. Kim, Asphaltene inhibitor testing: comparison between a high pressure live-fluid deposition and ambient pressure dead-oil asphaltene stability method, in: Offshore Technology Conference, Offshore Technology Conference, 2018.

[11] A.G. Di Lullo, T.P. Lockhart, C. Carniani, M. Tambini, A techno-economic feasibility study of asphaltene inhibition squeeze treatments, in: European Petroleum Conference, Society of Petroleum Engineers, 1998.
[12] L.M. Cenegy, Survey of successful world-wide asphaltene inhibitor treatments in oil production fields, in: SPE Annual Technical Conference and Exhibition, Society of Petroleum Engineers, 2001.
[13] K. Akbarzadeh, et al., Asphaltenes—problematic but rich in potential, Oilfield Rev. 19 (2) (2007) 22–43.
[14] S.-W. Wong, F. van der Bas, J. Groenenboom, P. Zuiderwijk, Near wellbore stimulation by acoustic waves, in: SPE European Formation Damage Conference, Society of Petroleum Engineers, 2003.
[15] L.C.R. Junior, M.S. Ferreira, A.C. da Silva Ramos, Inhibition of asphaltene precipitation in Brazilian crude oils using new oil soluble amphiphiles, J. Petrol. Sci. Eng. 51 (1–2) (2006) 26–36.
[16] K. Gharbi, K. Benyounes, M. Khodja, Removal and prevention of asphaltene deposition during oil production: a literature review, J. Petrol. Sci. Eng. 158 (2017) 351–360.
[17] M. Alhreez, D. Wen, L. Ali, A novel inhibitor for controlling Iraqi asphaltene problems, in: Environmental Impacts of the Oil and Gas Industries: Kurdistan Region of Iraq as a Case Study (EIOGI), 2017 International Conference on, IEEE, 2017, pp. 37–41.

Index

Note: Page numbers followed by *f* indicate figures, and *t* indicate tables.

G

H

I

K

L

M

N

O

P

Q

R

FSC
www.fsc.org
MIX
Papier | Fördert gute Waldnutzung
FSC® C083411

Kolibri 360 GmbH
Ferdinand-Jühlke-Straße 7
99095 Erfurt, Deutschland
produktsicherheit@kolibri360.de